BIZARRE BOTANICALS

BIZARRE BOTANICALS

How to Grow String-of-Hearts, Jack-in-the-Pulpit,
Panda Ginger, and Other Weird and Wonderful Plants

Larry Mellichamp
and
Paula Gross

Foreword by **Tony Avent**

Timber Press
Portland ‖ London

Frontispiece: Pink banana (*Musa velutina*) Photo by Larry Mellichamp.

Published in 2010 by Timber Press, Inc.

The Haseltine Building
133 S.W. Second Avenue, Suite 450
Portland, Oregon 97204-3527
www.timberpress.com

2 The Quadrant
135 Salusbury Road
London NW6 6RJ
www.timberpress.co.uk

Printed in China

Library of Congress Cataloging-in-Publication Data

Mellichamp, Larry.
    Bizarre botanicals : how to grow string-of-hearts, jack-in-the-pulpit, panda-ginger,
and other weird and wonderful plants / Larry Mellichamp and Paula Gross ; foreword
by Tony Avent. — 1st ed.
        p. cm.
    Includes bibliographical references and index.
    ISBN 978-1-60469-076-7
    1. Rare garden plants. 2. Plants, Ornamental. 3. Curiosities and wonders.
I. Gross, Paula. II. Title.
    SB454.3.R37M45 2010
    635.9—dc22
                                                            2010012442

A catalog record for this book is also available from the British Library.

From Larry:

To Audrey and Suzanne, the loves of my life

From Paula:

To my parents, who brought me to nature
To Alan, who reminds me that
"just being weird isn't enough!"

CONTENTS

Foreword by Tony Avent   9

Preface and Acknowledgments   11

Introduction   15

  1 Carnivorous Plants   31

  2 Ferns and Fern Allies   65

  3 Flamboyant Flowers and Bric-a-Bract   85

  4 Love Plants   117

  5 Odd Inflorescences   133

  6 Weird Leaves   151

  7 The Plant Zoo   167

  8 Prickly Plants   191

  9 Orchids   209

10 Succulents   239

Hardiness Zone Temperatures   274

Bibliography   275

Index   279

FOREWORD

⁓

BIZARRE BOTANICALS TAKES YOU on a fascinating botanical journey that combines the can't-put-it-down characters from a suspense novel with the science of a National Geographic special, all while mixing in the fun of a Mr. Wizard show. Both the young gardener and seasoned scientist will be amazed by these stories of sex, death, and other botanical wizardry, with subjects ranging from insect-eating plants to floral pyromaniacs, from corpse flowers to plants that reproduce asexually.

I first met Larry Mellichamp more than thirty years ago, mesmerized while listening to his talk on botanical curiosities. In the decades since, I've still yet to meet anyone else who has grown as wide a range of truly bizarre plants and possesses the diagnostic understanding of the science behind each behavior. This experience combined with Larry's childlike fascination, natural teaching instincts, and desire to share his knowledge make him the perfect person to undertake this fascinating project. He and Paula Gross have delivered an emotion-evoking book that will have you belting out phrases like, "Wow, cool," "I didn't know that," "Oh yuck," and "Far out."

Cobra Lily blossom. Photo by Paula Gross.

Cuddle up in a comfortable place (though this is probably not bedtime reading if your significant other is nearby) as Larry and Paula take you on a voyeuristic journey in which weirdness is not only celebrated but necessary for survival. If you're like me, you'll enjoy learning the dirty little secrets of some of the strangest plants that you'll ever know!

—TONY AVENT

PREFACE AND ACKNOWLEDGMENTS

WHAT A WONDERFUL TASK, to be allowed to write about plants that one has personally grown and which one thinks are among the most exciting in the world, bar none. That is what I realized when I first became involved in this project in 2008, after recovering from the initial shock of its potential scope. I had been referred to Timber Press by Tony Avent, an adventurous, cutting-edge nurseryman and promoter of weird and wonderful hardy plants at his Plant Delights Nursery in Raleigh, North Carolina. As I began contemplating, "What *is* a weird plant?" I looked around and realized that a great many of these plant curiosities were right in front of me, growing here at the University of North Carolina (UNC) at Charlotte Botanical Gardens. These were plants from around the world, tropical and temperate, that I had become fascinated with over the years while traveling, visiting private gardens, and exploring botanical gardens. As a professional botanist, I had been reading the literature on the plants for thirty-five years and knew about their unusual forms and behaviors and understood how they worked. I became excited over the opportunity to bring that information to the general public in the form of interesting thumbnail sketches and teachable lessons in botanical, horticultural, and environmental principles. I have grown every plant I have written about, so much of my observation comes from firsthand experience.

As I continued to ask myself, "What is a weird plant?" (a question with many potential answers), I began discussing this with my staff at the botanical gardens. I was thrilled when colleague Paula Gross, assistant director of the gardens, agreed to coauthor the book. Paula is also interested in unusual plants and the teachable moments they inspire. In addition, she enjoys taking plant photos and writes with a flare that added a new dimension to the project.

Carnivorous plants were a main impetus for this book, and so we have begun with them and allotted them the most space. We could have written twice as much about most of the plants in this book, as well as added numerous other plant examples. We encourage you to explore them further in the books listed in the bibliography, to find additional examples of weird plants, and to utilize the Internet to see many wonderful images and find more information on these plants. It is inevitable that we have left out some of your personal favorites, so we ask your forgiveness in advance!

We would like to thank the staff at the UNC Charlotte Botanical Gardens for their efforts at helping us decide on the merits of some of the weird plant suggestions, and for their tolerance of our distraction by the project. They take care of our weird and otherwise wonderful plants for the benefit of all who visit. Thanks go to Tammy Blume, Connie Byrne, John Denti, Teri Edwards, Meredith Hebden, Tina Lawing, and Sue Richards. Dana Harris contributed opinions as well. We received invaluable assistance from Paula's father, Richard Gross, in the form of the excellent photos that he generously provided.

We wish to thank the photographers who contributed their work. Each has been acknowledged with his or her photo. For photographic access to special plants we would like to thank Harold Blazier, Tom and Shan Nasser, Genie White, Atlanta Botanical Garden, and Daniel Stowe Botanical Garden.

We could not have progressed and persisted in this project without the help of our spouses: my Audrey and Paula's Alan. They had to endure some loneliness, likely became tired of hearing about "the book," and yet provided encouragement all the while. They share in the outcome, as it was a labor of love.

Finally we greatly appreciate the help we received from the top-notch staff at Timber Press, who encouraged us and answered our questions all along the way.

My University of Michigan graduate school mentor, the late Warren (Herb) Wagner, encouraged me to teach, and tropical biologist Daniel Janzen inspired me to celebrate the excitement of understanding plants. Paula's parents introduced her to the joys of nature early in life, and her professors and fellow students in graduate school at the University of Georgia horticulture program strengthened her connection to plants.

We hope you will be inspired to view plants in a different light after reading about their unusual behaviors. Some of the accounts in this book will make you laugh, some will make you cry, and most will make you wonder, *why?*

INTRODUCTION

How to Understand Weird Plants

WHEN YOU WALK DOWN A BUSY STREET and look at all the people going about their everyday lives, think about three groupings of individuals based on their livelihoods. The first and largest group includes the folks in the most familiar and plentiful professions, such as secretaries and salesmen, doctors and lawyers, teachers and truck drivers. All important, well known, and expected. We see them every day, and they are a part of our daily lives (we are them!). Those in the second group do something a little bit different. They are candy makers and gourmet chefs, inventors and novelists, Wall Street deal makers and sports figures. We see them occasionally. They are a bit outside our normal lives, and we may be intrigued to learn more about them, but they don't exactly shock or surprise us. The third group, the rarest, is not often encountered, but they are the group that excites us the most because they do extraordinary things that take special skills: magicians and acrobats, rock stars and daredevils, mountain climbers and astronauts. We are truly fascinated by their feats, and they fill us with a sense of wonder and imagination.

If you will accept, as I do, that we can view the world of plants as similar to the world of people in terms of diversity, then this book is

Titan arum (*Amorphophallus titanum*), perhaps the World's most fabulous plant. Photo by Connie Byrne.

mostly about the third group. Just as you find the most interesting people in strange places—the three-ring circus, the electrified stage, the nose cone of a rocket—you find the weirdest plants in some of the strangest places on earth and in the gardens of people who have a taste for the out-of-the-ordinary.

Why are Disneyland, the circus, and the roving carnival so much fun? Because they are such a change from our daily lives. It's the thrill of fast and furious rides. It's the bigger, stranger, more-colorful-than-life clowns, characters, and acrobats. It's the scantily clad, exotic ladies and the tunnel of love. It's the scary and forbidden haunted house and chamber of horrors. We love to experience the unexpected, the surprise of the unknown, the poof! of the magician's flash, the rush of discovery that may be a little bit dangerous. This desire to experience the macabre and grotesque, the ugly and inexplicable, the intricate and exceptional, burns

in all of us, no matter how small the embers, but rarely do we get to fan the flames and find out more or get directly involved.

Look at that perfectly normal houseplant there on your desk. It doesn't do much, just sits there, barely changing. Is that how you want to go through life? Why accept that this is the best the plant world can give you?

Anyone can find something of interest among odd and curious plants. After all, they may be strange, but they are still beautiful. Enjoy this tour among some of the most bizarre and unbelievable plants in the world. Fan the flames of your inner adventurous gardener, and grow a few of these plants yourself. We promise you will enjoy the ride.

Selecting the weird and extraordinary from the

Let your imagination go, and see the man-in-the-moon image in this unopened flower bud of the bucket orchid (*Coryanthes macrantha*). Photo by Paula Gross.

plant kingdom is a very subjective, personal endeavor. We chose to narrow our field of vision by establishing some guidelines: the plants featured in this book are all (well, mostly) naturally unusual (no cultivar monstrosities or one-of-a-kind plants from the wild), reasonably available for purchase, and reasonably growable. Meeting these three criteria eliminated some of the most famous weird plants in the world but still provided an outstanding field to choose from.

Our first guideline means that the plants we feature have naturally evolved the adaptations that we find so unusual. These traits and behaviors make them successful in their own, sometimes extreme, habitats. In

The normal growth form (erect, cylindrical stems) of this *Mammillaria* cactus, which can be seen in the pot, has been disrupted by a hormonal imbalance that has caused some of its stems to grow outward and produce a crest. This abnormal growth form will not come true from the seeds of this plant and can be perpetuated by asexual cuttings only. Photo by Larry Mellichamp.

Tips for Growing Your Own Weird Plants

Most of the plants in this book are not difficult to grow using standard techniques for the plant group. If you are a beginner, seek out experienced gardeners. There is no substitute for experience, and you're never too old (or too young) to start gaining your own. Here is a review of the major factors to consider when growing "weird" plants. Each plant description in this book also contains cultivation details, with the basics listed under "Growing Tips," including information about how difficult or easy the plant is to grow, whether it should be grown indoors or outdoors, light requirements, hardiness, moisture preferences, and ideal growing medium.

Light and light levels: Respect the light preferences of plants and you will save yourself a lot of heartache. If outdoors, does it need sun or shade? If indoors, does it need bright light in a sunny window or moderate light in a shadier window (or placement further back from a bright one)? Watch for leaf burn, caused by too much light or heat, and stretching or lack of flowering indoors, caused by too little light.

Temperature: "Hardiness" does not mean how tough a plant is. It refers to the minimum temperature a plant can endure without killing it (see "Hardiness Zone Temperatures" at the back of the book). Temperate-zone plants must be grown outside, as they need seasonal temperature changes to meet dormancy requirements. Many of the plants in this book are tropicals and prefer temperatures around 70°F (21°C) but can tolerate much higher. Most tropical plants go dormant at 50°F (10°C) and have "freeze" damage at 40°F (4°C). This is especially important to note if you will be growing them outdoors in the spring, summer, and fall and bringing them in for winter. The outdoor plants featured in this book are given a hardiness rating. Note, however, that even plants described as "not hardy" can be grown outdoors for the summer.

Water: Both overwatering and underwatering can cause problems. Just because a plant is doing poorly does not necessarily mean it needs more water! Check soil carefully—stick your finger in it, observe the color of the potting soil. Water thoroughly when you do water, and don't water again until necessary. This is easier said than done. Practice, and kill a few plants. You'll learn. Follow this adage from J. C. Raulston: "If you are not killing plants, you are not really stretching yourself as a gardener." If you are unsure of your water quality, have it checked or ask a local gardener. We use city water, chlorine and all, for most plants. However, we have found that some plants are sensitive to chlorine, especially ferns, some begonias, and some delicate tropical plants. If you are concerned, install a simple dechlorinator or collect nature's water in a rain barrel.

Soil: As with light, if you respect the soil preferred by your plants, you are well on your way to success. For an average indoor plant, start with a good-quality potting soil (don't bother with the cheap stuff). Do not allow it to get bone-dry or it will be difficult to rewet. We have given specific formulas in entries as appropriate for specific plants, such as succulents, orchids, and carnivorous plants. In containers as well as in the ground, good drainage is almost always recommended. If planting outdoors in poor soil, take the time and expense to improve your soil with compost, soil conditioner, or whatever amendment is appropriate for your soil type. Our advice on fertilizing: do it! All plants need nutrients, just not too much at once or at the wrong time of year (for some). Many indoor plants respond well to the "weakly, weekly" formula: 1–2 teaspoons of soluble fertilizer per gallon, used once a week. Any particular fertilization requirements are noted in individual plant descriptions.

almost every case these odd and curious plants would come true from
seed and perpetuate their behavior from generation to generation. They
are not mistakes of nature. They are just a little (or maybe a lot) different.
They're thrilling to watch, and that makes them all the more interesting.

The plants we decided not to include are those that are indeed true
freaks of nature—plants that produce abnormal growth structures due
to the influence of microorganisms, screwed-up genetic development,
and hormones gone amok. Among those excluded, for example, are
weeping trees; dwarfed conifers; many plants with variegated leaves,
flowers with color breaks, albino and doubled flowers, or seedless fruits;
and odd forms that are quite widely grown in cultivation. Such plants are
abnormal—monstrosities, aberrations. They are interesting but not natural. What is amazing is that these plants still grow and function but do
so only if protected and propagated asexually by humans. They would
not live very long in nature. They are found as lone individuals and propagated as ornamental plants to grow in gardens where they amuse those
who look for the truly unique. We enjoy these plants too but decided to
focus instead on the natural wonders the plant world has to offer.

To help you determine exactly how easy or difficult it is to grow the
plants featured in this book, we have assigned each a difficulty rating,
included in each plant entry under "**GROWING TIPS**":

1 Easy to grow. The plant should grow well if basic conditions are met,
with average soil, water, sunlight, and temperatures within the prescribed range. For houseplants, no dormancy or special treatments are
required. For hardy outdoor plants, average well-drained garden soil
within the prescribed light regime and zones. Proceed with no worry.

2 Relatively easy to grow. One especially critical condition must be met
for success, whether this means special soil type, watering regime, sunlight level, or dormancy requirement. Proceed with attention.

3 The most difficult to grow. Either more than one special condition
must be met, including a critical dormancy period; a series of conditions

must be met; or the plant just seems to have an inherent propensity for persnickety behavior. Proceed with caution.

Botanical Aspects

We offer this brief account of aspects of the plant body to help you understand the different forms and behaviors presented in the plant entries in this book. As you'll see, plants can modify their roots, stems, and leaves for various purposes—either temporarily or permanently through evolutionary changes known as adaptations. It is often these adaptations, which are just the result of nature working like it should, that make a plant "weird" to our human eyes.

Roots

The body of a plant is quite simple compared to that of most animals. Plants have only three (nonreproductive) organs: roots, stems, and leaves. Roots, like the analogies we draw from them, are foundational, supportive, often underappreciated. They are the first structure to emerge from seeds, grow toward gravity, and serve to anchor the plant and absorb water and nutrients. Roots are rarely seen but are crucially important and should not be unnecessarily disturbed.

In nature you can find roots that do unusual things, like the aerial roots of an epiphyte that grow around the branches of trees and dangle in the air. The thickened, spongy roots of some desert succulents can swell to store large amounts of water. We know that some roots can enlarge and store food because we eat them—carrots and beets, for instance.

Trash basket roots of the epiphytic orchid genus *Ansellia* grow upward and catch organic debris. Photo by Larry Mellichamp.

And finally, some roots can defy their own rule of growing toward gravity by growing against it, upward. This is the case with the "trash basket roots" of some epiphytic orchids and the "breather roots" of seaside mangrove trees. In the case of the orchids, the roots form a mass of upward-growing spikes, catching falling organic debris that accumulates and decays, releasing nutrients that feed the plants. Thanks to their special roots, these plants create their own little compost pile!

Stems

Stems are (normally) all about support, framework, and transport. They are the second organ to emerge from a seed and are connected to the roots through the internal water and food transport systems of the plant,

Desert tortoise plant (*Dioscorea elephantipes*) produces an unusual enlarged stem (caudex) with knobby bark. Photo by Larry Mellichamp.

the xylem and phloem. Stems provide the scaffolding for bearing leaves, flowers, and fruits. In the smallest of plants, stems can be fragile and soft, but in the largest, like trees, they must be incredibly strong, balanced, and able to add new growth each year, in girth as well as height. Stems follow the opposite gravitational rule as roots, growing upward, away from gravity. They also generally grow toward the light, as seen in the lean of plants toward a sunny window or the brighter edge of a shady forest. Stems have to be strong to hold up the heaviest fruits, and tree trunks must be structurally sound to hold up the leafy canopy.

Plants can't get up and move, but they can grow in a new direction, and stems are where new growth occurs. They are rather flexible in their development, adjusting their growth depending on the conditions they encounter, such as harsh weather, barriers they must grow around, other plants competing for space, and browsing herbivores. Scores of stem modifications are seen in nature, and stems can even take the place of roots and leaves if that is the pressure nature exerts upon them. Special stems can grow horizontally or twine around supports. They can enlarge and store food and water—potato tubers and water chestnut corms are actually stems. Stems can be modified into sharp thorns for protection or become short and thick for storage like the grotesque caudexes of some desert plants. For example, the stem tuber of desert tortoise plant (*Dioscorea elephantipes*) or Mexican yam (*D. mexicana*) produces bark in the form of large rhombic knobs, in which you can actually see the layering of yearly growth rings. Stems can entirely take over the responsibility of photosynthesis in plants that never bear leaves. The Mediterranean butcher's-broom (*Ruscus aculeatus*, pictured on page 204), for instance, has sharp-pointed stems that actually look like leaves but are anatomically stems, through and through!

Leaves

If stems can be flexible plant organs, so can leaves. Their primary mission, however, is food production through that amazingly powerful process called photosynthesis. Leaves are born on stems and vary greatly in shape and size as the plant ages or as the environment changes. Even on a single large plant, leaves in the sun can be smaller and thicker, while those in the shade are larger and thinner. The capture of light is only part of leaves' story, for they bear microscopic pores, or stomata, on their underside that allow carbon dioxide in (and oxygen out) for photosynthesis. In addition, the constant loss of water vapor through these pores is what actually pulls the water and nutrients up from the roots and through the body of the plant. This process is called transpiration, and the evaporation of water into the air serves to cool the plant (and its surrounding environment). Of course, too much water loss means wilting, although most leaves are quite resilient and will perk back up if water becomes available to the roots fairly quickly.

The leaves of Brazilian edelweiss (*Sinningia leucotricha*) have an unusually thick covering of soft, silky hair. Photo by Paula Gross.

Leaves have waxy coatings to help retard excess water loss and reflect extra sunlight (yes, they can get too much). Some go further and are covered with protective hairs that are even better at reflecting sunlight and cooling the plant, and that give the plant a fuzzy gray appearance. Leaves can be drastically modified as well. There are succulent leaves, enlarged for storing water. Spines are actually leaves that have evolved to become hard, sharp, and threatening to herbivores. Some leaves act as tendrils for climbing. Others may be colored to act as pollinator attractants. And believe it or not, evolutionarily speaking, leaves went through a series of changes to become flowers.

Flowers

Flowers are all about sex and are the showiest parts of plants. While the conspicuous ones aren't trying to attract each other (like animals do), they *are* trying to attract. They want to lure the creatures (birds, bees, butterflies, and so forth) that serve as messengers of their love, to carry pollen from one flower to another of the same species. Although flowers appear to come in endless variety, all consist of the same types of parts arranged in a predictable order. From bottom to top, the parts are *sepals*, usually green, to protect the unopened bud; *petals*, colorful to attract pollinators; *stamens*, the male part of the flower, bearing anthers that produce pollen grains that are dispersed by animals and wind, containing sperm cells; and *pistils*, the female part, consisting of a stigma to receive pollen, a style for the male pollen tube to grow down through, and an ovary that contains ovules with eggs. After fertilization, the ovary and ovules enlarge to become the fruit and seeds, respectively. Pollination and fertilization take place rather

Wax flowers (*Hoya carnosa*) have thick, long-lasting petals.
Photo by Larry Mellichamp.

quickly. However, it may take weeks or months for the fertilized ovules to develop into seeds and for the ovary to ripen into a fruit. Think of an apple: flowering and pollination for a few days in May, ripening fruits much later in October.

Even though flowers consist of the same types of parts, each one has a shape, size, color, sometimes odor, and flowering time that make it optimal for attracting and effectively utilizing the services of specific pollinators. Hummingbirds like red, tubular, odorless flowers; bees like fragrant blue and yellow flowers with a throat; butterflies like clusters of smaller, fragrant flowers; moths like white, tubular flowers that open with fragrance at night; flies like tiny flowers that smell bad; and bats like big flowers with lots of nectar available at night. Wind-pollinated plants have nonshowy flowers without colorful petals and produce copious amounts of pollen (leading to hay fever in humans) for widespread dispersal and hit-or-miss pollination. Flowers may offer rewards of pollen and nectar, or may lure insects by deception, gaining their services but not paying the bill, so to speak. Occasionally it is not the actual flower that does the attracting. This is where an optional, associated floral part comes into play: bracts. Bracts are modified leaves closely associated with flowers. They may be green and protective like sepals, or large, colorful, and flamboyant like petals. For example, the big, red, leaflike bracts of the poinsettia attract insects to the tiny, inconspicuous yellow flowers.

We tend to have a stereotypical view of flowers based on our common florist-shop examples. Each floral part can vary tremendously in its size, form, color, and sturdiness from plant to plant. We think of petals as being short-lived, but in some plants, such as many orchids or the wax flower (*Hoya carnosa*), petals can be very thick in texture, covered with a hairlike coating, and designed to last for several weeks. Sometimes the petals aren't showy at all, and the sepals or even the stamens carry the show. Flowers even display "behavior" of a sort in their interactions with pollinators. What do you expect when sex is involved?

Flowers vary greatly from species to species, but not *within* one species, meaning that a particular species of plant tends not to show variation in its flowers when it blooms from year to year (or through the scope of thousands of years). Flowers are very specific organs of sexual reproduction. They cannot change easily or it would disrupt the delicate process of pollination and fertilization, and the plants would not be able to reproduce. Because they are considered evolutionarily stable, the particulars of flower parts—the number of parts, their shape, arrangement, and other details—are used to define plant groups based on relationships. It is flower structure that determines a plant's family and relatives.

Plant Names

Dianthus floribus solitariis, corollis lacero-partitis, squamis calycinis ovatis acutis. That was the official Latin name for the common carnation in 1763 when Carl Linnaeus invented the system of binomial nomenclature used today. He wanted to replace such old, cumbersome, multiword names, which had been used for centuries. These "phrase names" were actually brief plant descriptions, and this one translates as "the dianthus with one flower, fringed petals, and round, pointed sepals." Linnaeus gave every known plant a two-part Latin name consisting of a genus name and a specific epithet (often erroneously referred to as the "species name"). In this case, the binomial of the carnation is *Dianthus caryophyllus.* What an improvement! Yes, it is still Latin and foreign to the ear and tongue, but it is also well within the grasp of the motivated plant lover.

Going a step further and breaking down the two-word name can actually be entertaining and educational. The genus name is a noun, representing a group of very closely related plants. The specific epithet is an adjective modifying the noun, and represents an individual kind, or species, within the larger genus group. So the genus *Quercus* refers to all oaks, a widespread group that is easy to recognize. *Quercus* is the word the people in ancient Rome used when referring to an oak tree, so we have adopted that name for oaks. Within the oak genus are many different species of oak: *Q. rubra* (red oak), *Q. alba* (white oak), *Q. velutina* (black oak), and so on. This would be analogous to a human family named Smith. Smith would be the genus, with specific members including Smith, Bill; Smith, Wanda; Smith, Joey; and Smith, Suzy. (We just turn the names around in our everyday naming of people.) Both parts of the Latin scientific name can mean something. They may be named after people (as in *Magnolia*, named after French physician and botanist Pierre Magnol), places, or structures. Or the names may reflect some distinctive characteristic of the plant—*Magnolia grandiflora*, for example, is a large-flowered magnolia. Also note that Latin plant names are always underlined when handwritten and italicized in printed works.

Many plants have common names. These are the names used by people down through the ages, in their native languages, for the plants in their domain. Sometimes a common name used by one person will be unknown or misleading to another person and is therefore not very precise. For example, gay-feather is a well-known common name for *Liatris spicata* (in the aster family, Asteraceae), but this species is also known as blazing-star. Blazing-star, however, is a common name for *Mentzelia decapetala* (in the loasa family, Loasaceae) in South Dakota. And in other western states, *Liatris spicata* is known as button-snakeroot, but not in eastern states. Confused yet?

The purpose of names is to facilitate communication and foster precision. We use common names in this book because, for casual readers at least, they are easier on the eyes, ears, and mouth. We also give the Latin (scientific) names to avoid confusion and so that you can effectively and accurately look up the plants elsewhere.

To name something is a beginning to understanding it and its place in the universe. To better understand plants, botanists further categorize plants by grouping them with related plants and giving these groups family names. We often give family names for the plants described in this book, as it can be an enriching experience to look beyond an individual plant to its relatives. Family names always end in -*aceae*. Asteraceae, for example, is a family name based on the most typical member genus, *Aster*. In this case, the common name is the same as the scientific genus name. This is not uncommon, and so you might recognize a good many genus names as well as the family names derived from them: Rosaceae (rose family), Liliaceae (lily family), Orchidaceae (orchid family). Wikipedia is a quick and reliable source of family designations. What would

Linnaeus think? I rather think he'd be proud that his system has become so accessible.

Resources

Every plant in this book is reasonably available, either from a commercial grower or online source. Some will be easy to find, others more challenging. Check with your local plant societies, botanical gardens, and knowledgeable people at garden centers. Experienced plant people love to share information and plant sources with each other. Start making those connections. When shopping online, be warned that some plants may not be what the seller says they are. Make inexpensive choices until you feel confident with your skills and the source of the rarer plants. But also don't be afraid to pay a reasonable amount for an unusual plant. They can last a long time and reward you much. Think about how long a twenty-dollar dinner lasts versus a twenty-dollar cactus.

Don't hesitate to consult the Internet for information, but do be warned that misspelled names and unclear queries can lead you down the wrong path. Sorting through all the junk and irrelevancies may prove frustrating, especially to beginners. Some good places to start are state cooperative extension sites, reputable nursery sites, and databases associated with botanical gardens and national plant organizations. Even though Wikipedia is open for anyone to edit, we have most often found its botanical information to be accurate.

Finally, look into the items listed in the bibliography. There is a wealth of new and old books on particular plant groups. These books often explain things well, and collectively have many more examples than we are able to offer. Reading these books can take you all over the world and into some strange places. They can give you ideas on what to look for and how to appreciate plants. If you are a young person, this will be the beginning of a lifelong source of joy and amusement. If you are an, um, experienced person, it will give you a deeper appreciation for the plants you have encountered and lead you on to new adventures.

1

CARNIVOROUS PLANTS

The Flies' Demise

IF ONE GROUP OF PLANTS is across-the-board weird and fascinating to the greatest variety of humans—young and old, plant-loving and plant-indifferent—surely it is the carnivorous plants, which "feed" on flesh. Widely grown, often misunderstood, they seem to defy explanation.

Why are they so universally intriguing? Because their lifestyles violate what seems like a fundamental rule of life on earth: plants don't eat meat—in fact, they don't eat at all! They make their own food and form the base of the food web. Everything else eats plants (or other creatures which themselves eat plants). A plant is for food and making oxygen, for creating habitats and providing fuel, fiber, and spices. Even if we don't often consider those "nurturing" qualities of plants, we at the very least think of plants as passive scenery. Carnivorous plant behavior is so fundamentally unplantlike, it flies in the face of reason (pun intended). No one expects a plant to turn the tables and actually capture and kill the insects that come seeking to graze upon its green bounty. A plant grabbing a bug and eating it? Agh! What's next? If a plant can eat a fly, maybe

A variety of carnivorous plants grow in the humid, bright environment of a large terrarium. Designed and grown by Martha Miller. Photo by Larry Mellichamp.

there is a plant out there that could eat me. Our imaginations run wild from there.

Even rational scientists with their fantasies in check are captivated by these amazing plants. Charles Darwin considered carnivorous plants fascinating, especially sundews and Venus flytraps, and studied them much of his life. In his 1875 book *Insectivorous Plants*, the first treatise devoted to these plants, he reported his detailed experiments on their movements in capturing and digesting prey. He found, for example, that a pebble would cause a flytrap to close but not seal tightly, and the plant would simply reopen the next day. Add some protein, however, like egg albumen, and this would trigger the tightening of the trap and the release of digestive enzymes. The plant could tell when it had caught a real meal.

So why does a plant that can make its own food by photosynthesis need to eat meat? The short answer is, to supplement its nutritional needs and allow it to compete and grow in nutrient-poor habitats. The next logical question is, "How does it manage to eat?" (with no teeth, stomach, or intestines). Before we reveal those secrets, let's take a quick look at who makes up this rogue's gallery of plants with a taste for prey.

More than 600 species of carnivorous plants have been recognized in the world, yet most of them can be divided into basically five different types—pitcher plants, sundews, butterworts, bladderworts, and the one and only Venus flytrap. Pitcher plants are the largest carnivorous plants and have tubular leaves that form pitfall traps. Sundews have variously shaped leaved that are absolutely covered in glistening, sticky hairs that act like flypaper, engulfing small insects in a mucilaginous mire. Butterworts have rosettes of broad, slimy leaves for insects to stick to. Bladderworts grow in water or waterlogged soil and bear many tiny, bladder-shaped leaves that quickly inflate to suck in miniscule prey. Because these leaves are so small and grow underwater or in the soil, they are not easily observed, and we do not cover them in this book. Last but not least is the international poster child for carnivorous plants, the Venus flytrap. Only one species is recognized from one area of the world, the coastal Carolinas. Its rapid-action and bear-trap-type leaves seem the ultimate in plant-turned-predator innovations.

Suppose a mild-mannered plant wanted to join this gang of insect-eaters. What does it take to earn the title of carnivorous plant? The "How to Be a Carnivorous Plant in Four Easy Steps" mantra sounds something like this: Attract. Capture. Digest. Absorb.

Attraction—Soon to Be Fatal

In order to attract prey, carnivorous plants appeal to one of two desires of insects: food, in the form of sugary sap, or a place to lay eggs, where the plants mimic carrion. While insects may visit carnivorous plants seeking sweet nectar, the flowers are not the source of this nectar, nor are they the capturing organ—those tasks belong to the leaves. Eating your *floral* visitors would be like shooting the messenger. The plants still need the services of insects to effect cross-pollination. Carnivorous plants act a little like Jekyll and Hyde. They allow their pollinators to visit when in flower at pollination time. But let the same bee come back later in the season to get nectar from the trap, and they'll be eaten without mercy or remorse. Carnivorous plants may attract prey by displaying color markings on their leaves that make the leaves resemble flowers, glistening nectar, or rotting meat.

Capture—The Autotroph's Revenge

If it is to qualify as "captured," the prey must come to the plant and be caught on purpose, not just by accident. Many noncarnivorous plants entangle or entrap small insects on sticky hairs on their stems, leaves, or flowers. But in these cases the plant is merely trying to protect itself from the hungry insects, not

Yellow pitcher plant (*Sarracenia flava*) and possibly suicidal fly. Photo by Larry Mellichamp.

purposely trying to capture, and this passive sort of trapping is of no direct nutritional benefit to the plant.

True carnivorous plants have modified leaves that act as clever trapping mechanisms, suited to particularly sized insects. The traps can be active, as in the snap traps of Venus flytraps or slow-moving glandular hairs of sundews. Or they can be passive, as in the pitfall traps of pitcher plants, in which there are no moving parts. Each method accomplishes the same end: captured and soon-to-be-dead insects.

Digestion—Eat What's Bugging You

Carnivorous plants, then, must do what animals do: break down their food into simpler molecules that can be absorbed. Some utilize specialized cells that secrete digestive fluids very similar to stomach acid. Others rely on the organic soup of dead victims, the decomposition action of microorganisms, and the body wastes of non-prey visitors to feed the plant. Unlike animals, though, the plants are not directly seeking the energy source of sugars. They are after the critical nutrients nitrogen and phosphorus.

Australian fork-leaved sundew (*Drosera binata* var. *multifida*) with various prey. Photo by Larry Mellichamp.

Absorption—Getting the Bugs Out of Your System

Finally, the plant must actually absorb the released nutrients. It does this via normal or specialized cells on the inner surfaces of the traps, much like the villi in human intestines that increase absorptive surface area. Absorbed chemicals can spread throughout the plant's body within a matter of hours. The extra nutrition these green plants acquire helps them thrive (not just survive) in nutrient-poor habitats where there is less competition from other plants.

Although numerous species of carnivorous plants belong to at least six different plant families and are found in various regions of the world, they live in habitats that share some particular features in common. Think of the medieval notion of the basic elements of earth, air, fire, and water. These are the factors carnivorous plants need to be happy. They normally grow in moist to wet, sunny habitats where there is high humidity (water and air). Even in such wet habitats, fire usually plays a major role in keeping the habitat devoid of dense vegetation. Throughout the southeastern United States, where many carnivorous plant grow, lightening-caused fires burn the wetlands periodically, and many plants are adapted to this regime (fire). The soils are usually low in nutrients either because they are acidic or because nutrients are leached by high rainfall. They are usually sandy with a lot of organic peat (earth). Under these conditions, if a plant can find nutrients somewhere besides the soil, it gains a real advantage over other plants.

Carnivorous plants grow in places that are often extreme and out of the way. One of the most famous locales is the Okefenokee Swamp in southeastern Georgia, a vast wetland of deep ditches and canals, dense alligator-infested swamps and marshes, and open water with floating islands of peat moss and rooted vegetation. Large clumps of the hooded pitcher plant (*Sarracenia minor*) can be found on the floating islands. Parrot pitcher plants (*S. psittacina*) and sundews grow thickly on the wet, sandy peat mats and ditch banks.

In the wetlands of Southeast Asia from Borneo to northern Australia, the vining *Nepenthes* pitcher plants grow in moist to wet, open but junglelike habitats where they climb on other vegetation using leaf tendrils. Their pitchers form at the ends of leaf blades and hang down like strange fruits suspended in midair, catching rainwater and providing reservoirs for all manner of tiny organisms. Some of these organisms serve as prey, and some actually live in the watery contents as members of a fascinating web of life contained in these miniature habitats.

In southwestern Australia, some 100 species of small sundews grow profusely in moist habitats, often where blinding white quartz sand provides a substrate. The famous western Australian pitcher plant (*Cephalotus follicularis*) can be found in marshlike grasslands or sandy seeps along beach cliffs.

The most diverse place on earth for carnivorous plants is the Green Swamp Preserve in southeastern North Carolina. There you will find a mosaic of wetland habitats marked by scattered hummocks of trees, deep black-water creeks, wet peaty bogs, and moist sandy meadows. The gradient of moisture and soil textures, combined with torrential rain and periodic fires, create the crucible in which carnivorous plants interact with other organisms and struggle for survival. Here grow four species of *Sarracenia* pitcher plants, four species of sundews, the only species of Venus flytrap, three species of butterwort, and about ten species of aquatic and terrestrial bladderworts. It is a magical place and is one of the last strongholds on earth for such plants to survive the onslaught of clearing, development, and fire suppression by humans. Some have estimated that as much as 75%–90% of their fragile wetland habitats in the southeastern United States has disappeared in the past one hundred years. The Green Swamp Preserve is a 13,000-plus-acre refugium.

All of these habitats produce surroundings that are uncomfortable to humans—hot, humid, mosquito-infested, mucky. You have to *really* want to go there to see and study them. Perhaps this has led to some of the mystery and misconceptions surrounding these plants. Explorers of old would come back from faraway and inhospitable places with partly factual reports of strange plants—or sometimes entirely fictional ones, such as the man-eating tree of Madagascar. In reality, no carnivorous plants in the world are big enough, or have the purpose or ability, to capture and eat anything larger than a small bird, lizard, or rodent.

The reputations of these plants have also been exaggerated by the man-eating plants of Hollywood movies, but like giant, angry ants and radioactive blobs with evil intentions, these plants simply do not exist. The Venus flytrap makes good material to draw from, but in reality its traps rarely get longer than about 1 in. (2.5 cm). The most famous carnivorous plant of first Broadway, then Hollywood, was Audrey II, the highly motivated potted plant in *The Little Shop of Horrors*. "Feed me! Feed me!" will forever haunt those who saw the play or the movie. Yet the clever rock-and-roll songs and choreography made the murderous man-eater somewhat endearing as well.

Whether in fantasy or real life, carnivorous plants capture human attention and fascination in a way that many other plants might only hope for (if plants have such human-attracting ambitions). Their allure may be deadly to insects, but for us they provide the vicarious thrill of life in exotic forbidding places, unconventional behavior, and the ruthless capture of prey—all lurking behind the striking colors and forms of what might otherwise be just another humble green plant.

WESTERN AUSTRALIAN PITCHER PLANT
Cephalotus follicularis

PLANT TYPE: Herbaceous perennial
HEIGHT AND SPREAD: Clumps 2 in. × 3 in.
(5 cm × 8 cm)

WE ARE NOT GENERALLY DRAWN to carnivorous plants because they are cute. But *Cephalotus follicularis*, the smallest of the true pitcher plants, is just that to some observers. Like a Chihuahua with an inner rottweiler, it is adorably miniature on the outside but vicious if you get too close.

This is the one species in the genus, and it occurs only in a narrow area of southwestern Australia from Augusta to Cape Rich. There it grows mainly in open, sunny, marshlike wetlands in acidic, sandy soils and on seepage cliffs. When I first saw this plant, I was hiking down a steep slope along an ocean inlet where the cliffs are partly formed from seams of coal, called Coalmine Beach. I gasped upon seeing it on the cliffs, as I had come so far for this very purpose. I was transfixed, looking at an array of 6 in. (15 cm) diameter clumps of perfectly formed, hairy, bloodred pitchers, each 3/4–2 in. (2–5 cm) tall.

Like almost all pitcher plants, the open mouth of the pitcher leaf is partially covered by a hood, which

GROWING TIPS

DIFFICULTY RATING: **3**
INDOORS VS. OUTDOORS: Indoors with humidity and temperature controls
LIGHT: Bright light
HARDINESS: Not hardy
MOISTURE: Keep constantly moist but not sopping wet
GROWING MEDIUM: 50% peat, 30% sand, 20% perlite
NOTES: Barely tolerates night temperatures above 70°F (21°C)

Did You Know?
You can get a cheap thrill by carefully inserting your finger into a Cephalotus pitcher. As you try to withdraw it, the plant's fangs will grip on, kind of like a botanical Chinese finger trap.

Western Australian pitcher plant (*Cephalotus follicularis*) growing on a seepage cliff in southwestern Australia. Photo by Larry Mellichamp.

keeps out most rainfall (so as not to dilute the digestive enzymes). These hoods appear almost striped, with translucent sections alternating with dark red sections. The most gruesome features of the pitchers are their teeth, which incurve like fangs over the rim of the pitcher mouth. You can get a cheap thrill by carefully inserting your finger into a *Cephalotus* pitcher. As you try to withdraw it, the plant's fangs will grip on, kind of like a botanical Chinese finger trap. These diminutive pitchers are remarkably tough in texture, and you can actually end up with a sore finger! Each pitcher has prominent hair-edged wings on its front exterior that may serve as a sort of ladder by which insects may crawl up to their doom.

The western Australian pitcher plant is among the most difficult carnivorous plants to grow well in cultivation, perhaps because it requires somewhat paradoxical conditions—a combination of very moist yet well-aerated soil; pure water and high humidity, but with air movement; and lots of sun, but not consistently high temperatures. If you would like to try your hand, purchase it from reputable dealers who can give you advice on growing these cute little killers.

COBRA-LILY
Darlingtonia californica

PLANT TYPE: Herbaceous perennial
HEIGHT AND SPREAD: Pitchers to 24 in. × 4 in. (61 cm × 10 cm)

Would you risk your life for a plant? William Brackenridge was the botanist on the first United States government expedition to explore the Northwest Territory from 1838 to 1842. In what is now northern California, near the base of Mount Shasta, he was allegedly being chased by hostile natives when he hurriedly scooped up a most unusual plant on his mad dash to make it back to his ship. The plant he risked pausing for was later named *Darlingtonia californica*, and those who have encountered it ever since have understood its allure.

Of course, the intrigue of visiting insects quickly turns to horror. The 2–3 ft. (61–91 cm) tall pitchers twist through more than 180 degrees as they rise from the ground. Each pitcher ends in a globose head, curving over so that its gaping mouth points straight down toward the ground. Hanging from the edge is a fishtail-shaped appendage, looking like the forked tongue of a serpent. To insects it may act more like a colorful flag, bearing droplets of nectar, enticing them up into the bowl of glowing light. The countless translucent spots among the green and red tissues illuminate the dome like a Crystal Cathedral (no salvation here, though). Lured inside, the insects find that the edges are slippery, and down they go. Coarse,

GROWING TIPS

DIFFICULTY RATING: **3**
INDOORS VS. OUTDOORS: Outdoors. North of zone 7, bring indoors for winter just before first frost
LIGHT: Bright light
HARDINESS: Zone 7
MOISTURE: Keep constantly moist but not sopping wet
GROWING MEDIUM: 60% peat, 40% sand
NOTES: Barely tolerates night temperatures above 65°F (18°C)

A cobra-lily (*Darlingtonia californica*) pitcher rears and twists above the ground, culminating in an enticing opening into a dazzling globe of light. Photo by Larry Mellichamp.

downward-pointing hairs line the pitcher's wall, impeding attempts to ascend. Cobra-lilies (they're not at all closely related to true lilies and are not to be confused with *Arisaema* species, also commonly known as cobra-lilies) produce their own pools of liquid inside, although surprisingly do not contain any plant-produced digestive enzymes. At the bottom of the pitcher, the condemned victim encounters a Dante's Inferno of bacteria, protozoa, and writhing fly larvae, all extremely efficient at turning insects into nutrient soup.

To see these plants in the wild, you would need to visit shallow streams and wet seeps from sea level to 8500 ft. (2591 m) in the Sierra Nevada of northern California and northward along the coast of Oregon. They form large colonies by their stoloniferous (runner-forming) growth habit. The best-known place is Darlingtonia Wayside along the coastal highway near the town of Florence, Oregon. The scene of an expanse of these plants in the wild is stunning.

Growing specimens in cultivation is challenging. Darlingtonias seem to need a good mimic of their native habitats—warm, sunny days but cool night temperatures below 65°F (18°C). Purchased specimens are probably worth trying once, in late winter, even if you can only observe these fantastic plants for the few months it takes them to realize they are not in cool northern California anymore.

VENUS FLYTRAP
Dionaea muscipula

PLANT TYPE: Herbaceous perennial
HEIGHT AND SPREAD: 3 in. × 3 in. (8 cm × 8 cm)

OH YES, THE WEIRDEST OF THE WEIRD, the pinnacle of carnivorous plantdom, the most famous plant in the world. How did this single species of a plant, no bigger than your hand and growing only in coastal wetlands of the eastern Carolinas, make it to worldwide stardom? Flashy character? Devious behavior? Great management and representation? Well, the flytrap *has* had its influential promoters—John Bartram, Thomas Jefferson, and Charles Darwin, who wrote of it in *Insectivorous Plants* and referred to it as "one of the most wonderful plants in the world." But unlike many rock stars today, the Venus flytrap doesn't need a lot of hype to draw attention. This star has made the world scene due to its own natural talents: colorful leaves that act like a bear trap, dramatic spines that look like gruesome fangs, split-second movements made in response to visiting insects, transformation into a green "stomach" complete with digestive juices that sometimes ooze from its margins. All *Dionaea muscipula* ever needed was a catchy common name.

Kids rarely fear the pitcher plants or sundews in

GROWING TIPS

DIFFICULTY RATING: **2**
INDOORS VS. OUTDOORS:
  Outdoors. North of zone 7,
  bring indoors for winter just
  before first frost
LIGHT: Full sun
HARDINESS: Zone 7
MOISTURE: Keep constantly
  moist but not sopping wet
GROWING MEDIUM: 60% peat,
  40% sand or perlite
NOTES: Best grown as part of
  a larger bog dish garden with
  other small carnivorous plants.
  Do not grow in enclosed
  terrarium in full sun

Did You Know?
Once Venus flytraps detect prey, they close so tightly that the trap becomes convex, forming a sort of stomach around the victim, to the point that it secretes digestive juices and can appear to actually drool.

Venus flytrap (*Dionaea muscipula*) grows close to the ground as a rosette of leaves with fatal attraction. Photo by Larry Mellichamp.

the bog plant displays at our university botanical gardens, but the Venus flytraps elicit a fear-excitement response. "Will it hurt if I put my finger in it?" "If I left it in there, would it dissolve my finger?" "How big do they get?" With this last question you can see their imaginations growing, and their hopes that maybe their pesky little brother might stumble into such a giant trap. Sometimes there is a little disappointment when they learn that the only creatures in danger are insects. But this does not last long, and most are eager to buy one and take it home for close observation.

The Venus flytrap grows as small rosettes or clumps of leaves (the stems are bulblike rhizomes underground). The actual traps are rarely more than an inch long, often with red coloration. Along the lobe margins are soft "teeth" that give the trap an extra margin of reach to ensnare those quick insects with hair-trigger reflexes for escape. On the inside of each trap lobe are three actual "trigger hairs," several millimeters long, that must be touched *twice* within twenty seconds (or two hairs touched) in order for the trap to snap shut in less than a second. This assures that nonliving disturbances do not cause the trap to close, expend energy, and get cheated out of a meal. If you make the traps close without feeding, they will reopen within twenty-four hours but are worse for the wear.

This gets at the mechanism of trap closure. Many scientists believe the closure action is caused by extremely rapid growth, where the cell walls suddenly soften and intracellular pressure causes the cells to grow faster on one side only, such that the leaf very rapidly "grows" itself shut. Then it must grow itself open, but this happens slowly over the course of a day. They can do this only so many times before they wear themselves out and no longer function. Once Venus flytraps detect prey, they close so tightly that the trap becomes convex, forming a sort of stomach around the victim, to the point that it secretes digestive juices and can appear to actually drool. Mmm.

After four to ten days, depending on the size of the prey, the trap reopens and the undigested husk of the insect lies crumbled against the wall of the trap. If a trap catches a particularly large insect, extra digestive juices may be secreted such that the trap leaves themselves are damaged and turn black. Similarly, if a trap is fed something rich in protein, like hamburger, it can die. To avoid this "indigestion," only feed your fly-traps freshly killed insects or small arthropods such as spiders, centi-pedes, and pill bugs.

Literally millions of Venus flytraps are grown every year from tissue culture and seeds to satisfy public demand. What kid (and often adult) has not tried to grow one? Yet many have not seen them live very long. It is not that they are especially difficult, but they do have some absolute requirements, and growing inside as a houseplant is not one of them. They must be grown in full sun, in an acidic medium, and kept always moist. Cultivation practices are based on mimicking the plant's natural habitat—that of the wet, low-nutrient pine savannas and bogs within a 60-mile radius of Wilmington, North Carolina. Historically the Venus flytrap had a slightly larger range, but it has been restricted by habitat loss due to drainage and development. Its natural habitat is fragile and dependent on constant moisture and frequent fires to burn off overgrow-ing vegetation. In the past, flytrap populations may have suffered from overcollecting. Currently U.S. law protects the plants in the wild, with collection punishable by fines of up to a thousand dollars per plant. Ironically, habitat can be drained and thousands of plants killed without any fines at all. But we do treat our stars, however much esteemed, paradoxically sometimes. Highly valued yet pushed to give us more, and sometimes the pressure is just too much. I hope the fascination and wonder inspired by growing the Venus flytrap will lead to greater curiosity about the universe of plants, weird and familiar alike, and bring renewed concern for preservation of wild species amidst our practice of wanton habitat destruction.

Venus flytrap (*Dionaea muscipula*) with wasp—a curious visitor soon to become an unfortunate victim. One of the darker trigger hairs can be seen below the wasp's wing.
Photo by Larry Mellichamp.

SUNDEW

Drosera

PLANT TYPE: Delicate herbaceous annual or perennial
HEIGHT AND SPREAD: Variable, typically less than 8 in.
× 3 in.
(20 cm × 8 cm)

IN THE NATURAL WORLD of carnivorous plants,
Venus flytraps are the country cousins, sundews the
global citizens. We're talking about distribution here.
The one species of Venus flytrap exists only in a small
area of the coastal Carolinas, whereas the more than
170 species of sundews are found on every continent
save Antarctica. Apparently the globally fashionable
way to be a carnivorous plant is to cover yourself in
big, sticky droplets of glistening glue. The diversity of
sundews is especially rich in Australia, Africa, and
South America. Whether in the ditches of the coastal
southeastern United States, the bogs of Europe and
Asia, the sandy wetlands of Australia and South Africa, or the high-mountain tepuis of South America,
sundews are seeing to it that carnivorous plants are in
every suitable niche on the planet.

Close up, the sundews reveal their strange appearance: tentacled, glistening with slime, unplantlike in
form, and able to move with slow but sinister action.
Insects attracted to the shimmering droplets and colors of sundews quickly find themselves in a sticky

A sundew (*Drosera capillaris*) with a dense rosette of
leaves grows close to the ground. Photo by Paula Gross.

situation. Leaf blades are covered in hairs tipped with droplets of sticky, non-water-soluble mucilage that lights up in the sunshine (hence "sundew"). Insects are caught in the natural glue, and the more they struggle, the more they get covered in slime, sealing their fate. The tentacles in contact with the prey detect proteins and actually move, pressing the insects close to the leaf surface. Shorter hairs secrete digestive enzymes and absorb nutrients. There is even one species, *Drosera glanduligera* in southern Australia, which has snap tentacles, whose outer hairs move very rapidly when touched to capture prey. Perhaps this or another snap-tentacle species, now extinct or as yet undiscovered, gave rise to the rapid movement in an ancient relative of Venus flytrap.

With so many species, one naturally expects variation in form and

size. The leaf blades come in many different shapes: round, oval, forked, spoonlike, and linear. The largest species is *Drosera regia* from South Africa, whose leaves may reach 3$^{1}/_{2}$ ft. (1 m) long. The smallest are the pygmy sundews of southwestern Australia, which when mature can easily be covered by a dime. Perhaps the most bizarre is the fork-leaved sundew of Australia (*D. binata* var. *multifida*), whose leaf blade can be repeatedly divided into as many as twenty-seven points, like the antlers of a gigantic buck. All species have showy flowers that can be white, pink, orange, or red. An additional unusual feature of sundews is that their leaves and flower stalks uncoil like a fern fiddlehead or a scorpion's tail—a rare trait in flowering plants.

With so many species available for purchase, it is difficult to give specific prescriptions for growing. Purchase sundews from a grower that specializes in carnivorous plants, and ask for his or her advice on specific soil mixes and seasonal care. The tropical species tend to be the easiest to cultivate, as they can grow continuously year-round without the need for dormancy. A good species to start out with is *Drosera capensis*, the cape sundew. Unlike many other carnivorous plants, tropical sundews can

grow well in a terrarium because they love high humidity and do well in moderate but not high light—perfect ambassadors for the global family of sundews. Some hardy species produce tightly compact overwintering buds called hibernacula (what a great word—sounds like a supervillain!) at ground level; others may behave as annuals.

Drosera peltata, the shield sundew, is a tall species from eastern Australia with tentacle-laden leaf blades born along an upright stem to 12 in. (30 cm) tall. Photo by Larry Mellichamp.

MARSH PITCHER PLANT
Heliamphora

PLANT TYPE: Herbaceous perennial
OTHER COMMON NAME: South American pitcher plant
HEIGHT AND SPREAD: Clumps 6 in. × 6 in.
(15 cm × 15 cm)

GROWING TIPS

DIFFICULTY RATING: **3**
INDOORS VS. OUTDOORS: Indoors
with humidity and temperature
controls
LIGHT: Bright light
HARDINESS: Not hardy
MOISTURE: Use pure (unchlori-
nated) water to keep moist
but not wet. Likes high
humidity
GROWING MEDIUM: 50% peat,
50% perlite
NOTES: Barely tolerates night
temperatures above 70°F
(21°C)

IN THE 1840S THE FIRST EUROPEAN explorers finally reached the remote, flat-topped sandstone mountain formations called tepuis in the Guiana Highlands of southern Venezuela. These scattered, mesa-like formations can rise to 3000 ft. (914 m) or more above the surrounding territory, and are effectively like islands. Such habitat isolation usually brings about the existence of many endemic plants and animals—creatures found only from those single, isolated locations. The mist-shrouded tepuis are home to many such endemic plants, and some of the first to be collected by European explorers were the *Heliamphora* pitcher plants. Since that time at least fourteen species have been described, with the anticipation of more to be found on yet unexplored tepuis. How they arrived, how they are related to *Sarracenia* of North America, and how they evolved on these high-altitude, cold-but-sunny, rocky, treeless, rain-drenched, marshy savannas, no one knows. They are cloaked in mist and mystery.

Most heliamphoras have particularly handsome, elegant, and well-proportioned forms. The graceful lines of the pitchers, the curve of their openings, and their coloration are reminiscent of beautiful blown-glass sculpture. Their hoods are greatly reduced into little, cupped, knob-like structures called nectar spoons. These are often red and secrete nectar, attracting prey. The pitchers have wide, sloping openings to admit insects, and many downward-pointing hairs to prevent escape. As the hoods are not large enough to keep out rainwater, many species have a special notch partway down their pitchers. This serves as a sort of over-flow valve to drain excess water out of the pitcher before it spills over

Marsh pitcher plant (*Heliamphora*) forms clumps of hoodless pitchers with distinctive, red "nectar spoons" that attract prey. Photo by Larry Mellichamp.

the top, carrying the nutritious insect juices with it. These species do not produce their own digestive enzymes, relying instead on the fungal and bacterial soup effect to decompose prey and make available extra nitrogen and other nutrients. The high rainfall of the area leaches nutrients from the thin soil. The ability to derive extra minerals from animal prey gives heliamphoras and the sundew species that grow there a survival advantage.

Heliamphora (like *Darlingtonia* and *Cephalotus*) is not easy to grow in cultivation, requiring those elusive cooler nighttime temperatures, bright light, high humidity, and yep, you guessed it, always moist, but well-aerated, freely draining soil. Heliamphoras are less readily available than most other carnivorous plants, so you will need to search a little harder to acquire one. But then, so did (and still do) the botanical explorers of the mist-shrouded tepuis of the Guiana Highlands. The fastest way to get there these days is by helicopter—or if you prefer, a three-day overland hike and then up a 3000 ft. (914 m) vertical cliff!

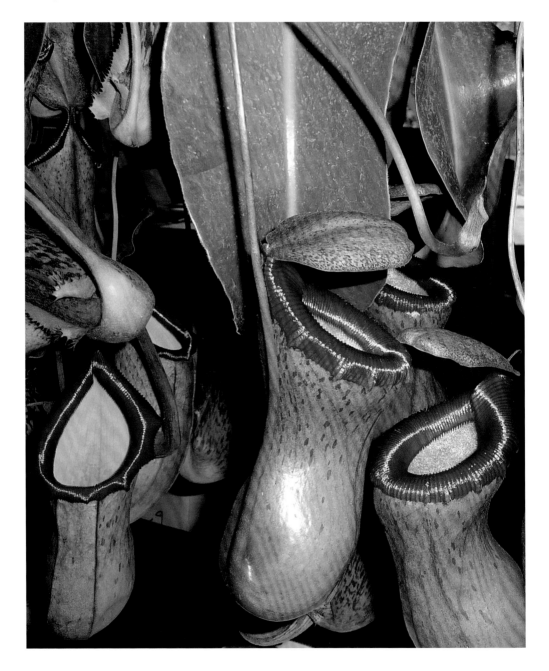

TROPICAL PITCHER PLANT
Nepenthes

PLANT TYPE: Woody perennial vine
OTHER COMMON NAME: Monkey-cups
HEIGHT AND SPREAD: To several feet long × 6–24 in.
  wide (one or more meters long × 15–61 cm wide)

PERHAPS THE MOST DRAMATIC-LOOKING of the carnivorous plants, tropical pitcher plants are masters of climbing, scrambling through the steamy wetlands of Southeast Asia bearing multitudes of strikingly colored vessels of death along their vine-like stems. The 100 or so species range from Madagascar through Malaysia to northeastern Australia. The greatest numbers are in Sabah, Borneo, and they are especially abundant on Mount Kinabalu.

These plants are found in moderately sunny areas like seepy hillsides, stream banks, and less dense jungle edges, as well as more open roadside wetlands. As these habitats do have some competing vegetation (fire does not play a role), tropical pitcher plants gain support and a spot closer to the light by climbing over and up into other plants. The leaves along the stems have a normal flat blade, but the tip is prolonged into a thin, wirelike tendril that coils around branches for support, in the same manner as the tendrils of grape vines. Eventually the tip of the tendril enlarges

GROWING TIPS

DIFFICULTY RATING: **2**
INDOORS VS. OUTDOORS: Indoors
  with humidity controls
LIGHT: Bright light
HARDINESS: Not hardy. Select
  species that will tolerate the
  high temperature extremes of
  your growing conditions, such
  as *Nepenthes sanguinea* or
  N. ventricosa
MOISTURE: Keep constantly
  moist but not sopping wet.
  Likes high humidity
GROWING MEDIUM: 50% peat,
  25% perlite, 25% fine orchid
  bark

Did You Know?

There is also an amazing food web inside the pitchers, with vicious and innocuous animals living there unharmed, feeding on prey, and in turn helping feed the plants—their homes.

Tropical pitcher plant (*Nepenthes*) hybrids with striking, large, thick-textured, dark-colored pitchers. Photo by Larry Mellichamp.

abruptly into a hollow pitcher with a lid. Some pitchers are impressively large, such as those of *Nepenthes rajah*, which can be 14 in. (36 cm) tall by 11 in. (28 cm) broad and hold a quart of water! When I was in Sabah some years ago, I drank a pint of fresh rainwater from a newly opened *N. rajah* pitcher (and yes, I did look inside first to make sure it contained no dead mice or insects). The genus name is from the Latin *nepenthe*, which refers to an ancient elixir that supposedly made men forget their troubles. I can't say I will verify that power exactly, but having the experience of drinking from that pitcher, and wondering how I would feel afterward, was a thrill I will always remember.

The pitchers of most *Nepenthes* species have dark red or blotched coloration that may be mimicking the appearance of carrion. The rims are slippery, and some have spinelike edges that help deter escape. Nectar is produced from below the hoods as an additional attractant. All species hold some amount of rainwater inside the pitchers for drowning and digesting prey. Some of the larger species have actually been found with small rodents, birds, and amphibian corpses inside. Still, insects make up the majority of their capture. There is also an amazing food web inside the pitchers, with vicious and innocuous animals living there unharmed, feeding on prey, and in turn helping feed the plants—their homes.

Coming from tropical climes, nepenthes are not winter hardy and so must be grown in greenhouses or sunrooms, and they take up a bit of space because they are climbers. Cut the vines back to near the base every year. This helps control size and stimulates new growth, producing the largest pitchers. Normally you would keep a small amount of water in the pitchers, and regularly feed them crickets, cockroaches, grasshoppers, and other meaty insects. Sometimes at night I turn on the lights in the greenhouse and go cockroach hunting. The large insects go into the pitchers. If I capture too many at once, I put them in a plastic bag in the freezer and dole them out weeks later. Weird behavior, I know—maybe that drink from the *Nepenthes rajah* pitcher had an effect on me after all.

BUTTERWORT
Pinguicula

PLANT TYPE: Herbaceous perennial
HEIGHT AND SPREAD: Flattened rosette to 2 in. × 3 in.
(5 cm × 8 cm)

HOW DOES AN INSECT-DIGESTING PLANT get a
common name like butterwort? Respected authors
seem to disagree on the meaning of the names. *Pinguicula* is the genus. The Latin *pinguis* means "fat," or
some have translated it into "little greasy one"—not a
strict reference to butter, though. The common name
of butterwort may have arisen on its own, simply

GROWING TIPS

DIFFICULTY RATING: **2**
INDOORS VS. OUTDOORS:
Outdoors. North of zone 7,
bring hardy species indoors
for winter just before first
frost. Tropicals never below
40°F (4°C)
LIGHT: Bright light
HARDINESS: Zone 7
MOISTURE: Keep constantly
moist but not sopping wet.
Prefers slightly higher than
average humidity
GROWING MEDIUM: 50% peat,
50% sand or perlite

The butterwort *Pinguicula ehlersiae* ×
P. moranensis is a free-flowering
hybrid that is easy to grow. Photo by
Paula Gross.

from the appearance of the leaves of European species. German monk and botanist Vitus Auslasser referred to the plant as *Zitroch chrawt*, roughly meaning "lard-herb." The leaves do look a bit like slices cut from something smooth and greasy, like lard or butter. What looks like grease from a distance is actually sticky mucilage, produced on small hairs covering the surface of the pale green leaves. The true nature of these hairs is revealed when you see flies (usually gnats) in the butter.

About eighty species of *Pinguicula* are scattered across the globe, in both temperate and tropical regions. While their seasonal behavior and appearances may vary, they all exhibit similar forms and flower structures. Butterworts are low-growing, rosette-forming plants usually measuring 1–7 in. (2.5–18 cm) in diameter. At certain times of year the rosettes produce showy flowers on tall stalks that greatly exceed the height of the sticky leaves. These are almost always pleasingly asymmetric in form (much like the flowers of violets) with long nectar spurs and variously colored petals of purple, pink, yellow, or white, depending on the species. But the carnivorous action happens on the leaves, of course. What seems merely greasy to our touch is actually quite sticky to small insects landing on the glistening leaf surfaces. Butterwort leaves respond to the presence of prey by secreting more mucilage, with which they further ensnare their victims. It has been shown (through radioactive tracing) that nitrogenous compounds from insects can be transferred to other parts of the plant in a matter of a few hours.

Butterworts often have delicate root systems, and many of them, especially in the temperate zones, go dormant in winter and produce hibernacula. These are more difficult to grow in cultivation. Tropical types can grow year-round in a sunny window, greenhouse, or terrarium under artificial lights—most are not exacting as to requirements. Many species and hybrids are available from select growers, so if you have the collector gene, this may be the carnivorous plant for you. Although butterworts are not at all threatening-looking and have an arguably silly common name, they turn out to be very efficient trappers and digesters indeed. You always have to watch out for the quiet, unassuming ones, don't you?

NORTH AMERICAN PITCHER PLANT
Sarracenia

PLANT TYPE: Herbaceous perennial
HEIGHT AND SPREAD: Clumps 2–24 in. × 4–8 in. (5–61
cm × 10–20 cm)

ON SEEING THE PITCHER PLANTS in the display
bog beds at our university botanical gardens, first-
time visitors often remark, "What are those weird
flowers? I've never seen anything like that." To which
we relish replying, "Why, those are not flowers at
all—they're leaves! Not just any leaves, either. They're
tubes of death." And just to prove it, we sometimes
slit one lengthwise to proudly display the stacked-
up masses of dead insect parts—the remains of a
season of meals for that pitcher—adding, "You
weren't the only creature fooled into thinking those
were flowers."

Sarracenia includes eleven species, all native to the
southeastern United States, and one of which also extends its range far into the north. The showiest is the
white-top pitcher plant, S. *leucophylla*, from the Gulf
Coast, so often mistaken for a flower. The top one-
quarter of its 2–3 ft. (61–91 cm) tall pitchers appear
like white and dark pink lace topped by a frilly mar-
gined hood. It is so darned attractive that its lure has
extended beyond its intended insect audience—all
the way to the floral industry. Rather than remaining
in the meadows capturing prey, pitchers are cut and
sold, later finding themselves captives in deluxe floral

GROWING TIPS

DIFFICULTY RATING: **2**
INDOORS VS. OUTDOORS:
Outdoors
LIGHT: Full sun
HARDINESS: Zone 7, or colder
with protection
MOISTURE: Keep constantly
moist but not sopping wet
GROWING MEDIUM: 60% peat,
40% sand
NOTES: Very heat tolerant

Did You Know?
*As the victims accumulate
in the depths of the pitchers,
digestive juices are secreted
and liquefy the soft bits for
absorption.*

White-top pitcher plant (*Sarracenia leucophylla*) leaves look like flowers and are used as "cut flowers" in arrangements. Photo by Larry Mellichamp.

arrangements, often in dried form. Martha Stewart has knighted them with her good favor by a spread in her magazine. Generally these florist-shop "cut flower" (leaf!) specimens are collected by knowledgeable owners who harvest from managed pitcher plant bogs in Mississippi and Alabama. The biggest threat to these Gulf Coast residents, both plants and people, continues to be the destruction of their habitats due to drainage and development.

Pitcher plants are found in sunny, wet, often peaty and sandy habitats in pockets throughout the southeastern United States, such as bogs, swamps, moist pine savannas, and fens. One species, *Sarracenia purpurea* (purple pitcher plant), ranges far beyond the Southeast. It is very widely distributed and extends so far north (zone 2) as to be almost arctic. It thrives in northern bogs that began as small lakes formed during the retreat of the glaciers 12,000 years ago. As the lakes aged, they completely filled in with sphagnum moss, which partially decomposed and formed large, mat-like deposits of what we call peat. Sphagnum peat bogs are sunny, very acidic, and low in nutrients—the perfect formula for carnivorous plants. In addition to the purple pitcher plant, sundews, bladderworts, and butterworts grow on or near these lake-margin mats, as do some noncarnivorous plants such as orchids and cranberries.

All sarracenias have tubular leaves, most with a "mouth" opening upward and covered by a hood to keep out rainwater and prevent winged prey from

Purple pitcher plant (*Sarracenia purpurea*) with gaping mouth, rainwater-filled tube, and erect hood. Photo by Larry Mellichamp.

flying out. The exceptions include the purple pitcher plants, which have gaping mouths with hoods held erect. All species produce nectar around their mouths and on their hoods, and have slippery surfaces and downward-pointing hairs lining the pitchers. As the victims accumulate in the depths of the pitchers, digestive juices are secreted and liquefy the soft bits for absorption.

The array of invertebrates that can be found in the tube of a pitcher at the end of a season is astounding. In one study, some 250 species were identified within the yellow pitcher plant (*Sarracenia flava*) after a summer of growth, including insects, spiders, and other arthropods. Early in the season, insects come seeking the sweet nectar secreted around the mouth. As these bugs die and begin to smell like rotting meat, they attract carrion-feeding insects. These in turn fall in the pitchers and die, adding to the stink. Next come the ambush bugs, seeking to catch their own live prey. They hang around and are often caught themselves. Occasionally you will see a small tree frog sitting in the mouth of a pitcher

Sarracenia tubular leaf cut open to show a season's accumulation of indigestible insect parts. Photo by Paula Gross.

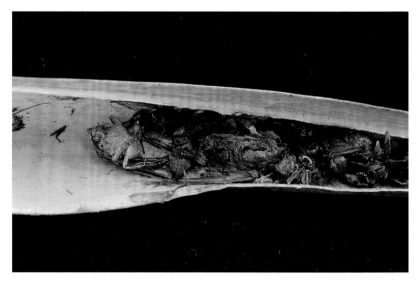

plant, catching the food the plant attracts and leaving a little organic fecal matter in the tube—a relationship benefiting both parties.

The true flowers of sarracenias are as unique and peculiar as the leaves. Each large, solitary flower hangs down on a tall stalk, shedding pollen into an upside-down-umbrella-shaped style. Bees enter the flower, wallow in the pollen, exit, and fly to the next flower. Successful cross-pollination results in a ³/₄ in. (2 cm) spherical seedpod with 500 or more seeds. When two or more species grow together in nature, they often hybridize, creating an array of offspring with intermediate forms and colors. When grown together in cultivation, all the species can hybridize, and many man-made hybrids of outstanding vigor, form, and color are available for hobby growers.

If you want to try your hand at growing a showy carnivorous plant, sarracenias are a great place to start. The plants are easy to grow in anything that can hold moist peaty soil and still be well draining. You can create an outdoor bog garden using a rubber pool liner inside a 12 in. (30 cm) deep bed of wall blocks. The best option for a smaller space is a 12 in. diameter dish garden in a plastic bowl with a drainage hole. A pitcher plant, sundew, flytrap, and bog orchid fit nicely in such a space and form a minibog garden. The garden must be outdoors in full sun and kept moist. We water every day in the hot summer. Pitcher plants go dormant and die down in winter. Cut the old pitchers off before spring growth begins to allow for the fresh pitchers to shine—like the flowers they aren't.

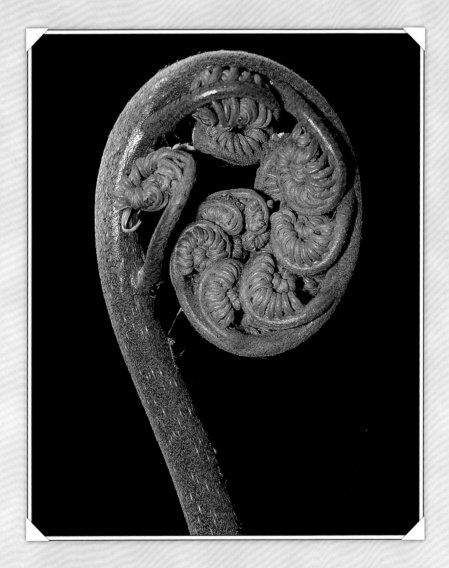

2

FERNS AND FERN ALLIES

With Fronds Like These, Who Needs Anemones

THERE IS SOMETHING COMFORTING about a fern. Like a favorite great aunt or uncle, they reassure us that while our lifestyles are different and perhaps more complex, there is something to admire, even envy, in age and staying power.

In the plant kingdom, ferns have staying power. They have been on earth for more than 350 million years, according to the fossil record. Long before the dinosaurs of Jurassic Park, vast forests of primitive plants, including ferns, covered the warm, moist earth. When these plants died and fell into the muddy swamps, they formed great beds of layered organic matter. Over the ensuing million years those deposits were compressed and heated by the earth and have become converted into the coal we mine today and burn as "fossil fuel." Hence, the period between 350 and 290 million years ago is called the Coal Age, a time when ferns and their relatives dominated the earth.

What makes a fern a fern? The short answer is fronds, fiddleheads, and free-ranging spores. Fronds are the relatively large leaves of ferns, usually divided into many smaller leaflets. We recognize an average fern by this feature, even calling other plants "fernlike" if they have delicate, much-divided leaves. Each fern frond starts out as a fiddlehead and

Uncoiling fiddlehead of mule's-foot fern (*Angiopteris evecta*) shows a primitive trait. Photo by Richard H. Gross.

uncoils as it develops. Ferns reveal their ancient origins and differ from conifers and flowering plants by not having seeds: they reproduce solely by spores. These are produced by the thousands (or millions) on the undersides of the fronds in clusters called sori. Fern spores are so small and lightweight that they have been found in air samples in the upper atmosphere of earth! So why aren't ferns taking over the world? It basically boils down to water. Environmental conditions have to be just so in order for fern sperm and fern eggs to create that magical moment together—sexual reproduction. A thin film of water is required to transfer swimming sperm from one tiny fern plant (called a prothallus and no bigger than half your pinky nail) to another. And all this occurs in the open environment, not protected inside the developing "seed" as in flowering plants and conifers. Although ferns grow from the tropics to the cold temperate zones, they are generally restricted to shady, moist, humid conditions where they periodically get that thin film of water hanging around long enough for a successful sexual union. Don't get me wrong—15,000 species of true ferns on earth ain't too shabby. Here we highlight a few ferns that stand out from the crowd (like that great aunt who wears pink spandex, has a Facebook page, and drives a Mustang).

MOTHER FERN
Asplenium bulbiferum

PLANT TYPE: Herbaceous perennial
HEIGHT AND SPREAD: 12 in. × 12 in. (30 cm × 30 cm)

MOTHER FERN IS NOT ONE of the weird plants that gets spotted across the greenhouse and makes you ask, "What is *that?*" No, it appears as a normal, feathery green fern until close inspection. Looking carefully, you realize something strange is going on with these fronds. They appear fluffy because they are covered in perfect little baby ferns.

Sometimes it is not enough that a plant can produce an abundance of seed, or in the fern's case, millions of spores. Some plants insist on giving "live birth" to genetic clones of themselves. *Asplenium bulbiferum* and the hybrid *A. ×lucrosum* exhibit this botanical phenomenon with the rather intimidating name of "pseudovivipary." It is analogous to a lower animal bearing live young without the sexual union of eggs and sperm (not at all common!). It is more like asexual plant runners or offsets. The little fernlets grow directly from the mother frond tissue, and if you look closely you can see their tiny, pinhead-sized fiddleheads. They will drop off and grow nearby into full-sized adult ferns.

So the mother fern has accomplished a form of reproduction that gets more of itself out into the local environment. But asexual (clonal) reproduction without effective dispersal is not generally thought of as the best long-term strategy for a plant. Sexual repro-

GROWING TIPS

DIFFICULTY RATING: **1**
INDOORS VS. OUTDOORS: Indoors
LIGHT: Bright shade
HARDINESS: Not hardy
MOISTURE: Keep evenly moist. Likes humidity
GROWING MEDIUM: Good, well-draining potting mix

Mother fern (*Asplenium bulbiferum*) produces little plantlets along its leaf that can fall off and take root. Photo by Diana Bradshaw.

duction assures the coveted and adaptive "genetic diversity" within a population, and dispersal gets the "teenagers" out of the house and into the world to find their own fortunes. These plantlets are genetic clones and grow near the mother, but they must be useful under certain environmental circumstances to crowd around the home front. While the phenomenon is rare, it occurs in enough species that it can't be dismissed. I am sure the mama fern wouldn't judge her little clonal fernlets as anything but perfectly normal.

MOSQUITO FERN
Azolla caroliniana

PLANT TYPE: Herbaceous perennial
HEIGHT AND SPREAD: Mat-forming individual plants
1/2 in. × 1/2 in. (1 cm × 1 cm)

SAY "FERN" AND MOST FOLKS form a mental image of a clump of arching green leaves divided into many delicate, ferny segments. Most people also realize that ferns like it moist (Grandma sternly reminds us of this when we forget to water her ferns while she is on vacation). Well, a few ferns take their love of water to the extreme, actually living in (on) water! That's right: aquatic ferns. Before you start picturing Boston ferns floating elegantly across a lake like an eyrar of swans, let me warn you—these are not the friendly fronds you are used to seeing. It might take some convincing for you to believe these tiny floating plants are actually ferns, and often vicious ones at that.

There are several genera of water ferns, and some are noxiously invasive weeds in aquatic systems, reproducing uncontrollably when conditions are right. The genus *Azolla* contains seven species and is distributed worldwide. Most are not hardy in areas colder than zone 8, but it is always better to be safe than sorry when it comes to potentially weedy or invasive plants (they seem to have a supernatural will to thrive). Do not let them go into the wild!

Carolina mosquito fern, *Azolla caroliniana*, is an often-grown species native to the swamps and marshes of the southeastern United States. This tiny, even

---

GROWING TIPS

DIFFICULTY RATING: **1**
INDOORS VS. OUTDOORS:
Outdoors. If growing a tropical species, bring a small amount indoors to a sunny window for winter
LIGHT: Full sun to bright shade
HARDINESS: Zone 7, depending on species
MOISTURE: Aquatic. If grown in soil, keep soil wet
GROWING MEDIUM: Grows on the surface of water (or wet soil) with no special requirements
WARNING: Some azollas are invasive in aquatic ecosystems. Never dump any water ferns into lakes or waterways. *Azolla pinnata* is a worldwide invasive plant, on the U.S. Federal Noxious Weed List, and should never be grown

---

Did You Know?
The thick carpet of water fern makes it difficult for mosquito larvae to get to the surface for air.

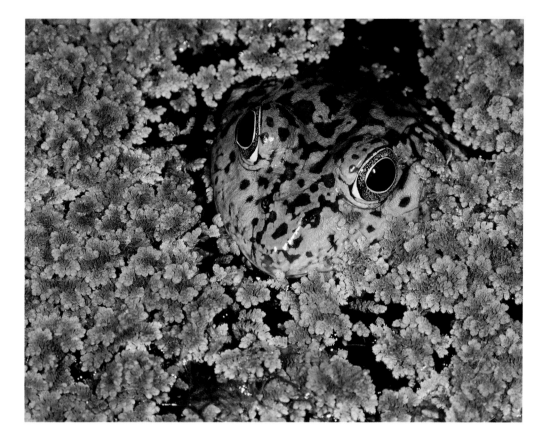

cute, plant is decidedly unfernlike in appearance and habit. Each plant consists of a series of overlapping, scalelike leaves, with fine roots hanging down into the water. Each little scale is 1–2 millimeters long. Toward the end of a season or in especially high light and low nutrients, this water fern rolls out the red carpet, covering the surface of the water like burgundy velvet. The plant is actually producing an abundance of anthocyanins (reddish pigments) in response to stress, but it sure looks pretty. Grow and appreciate this fern in its own individual water container, as it can take over a mixed water garden very rapidly.

There are layers of richness to the story of azollas, and perhaps their most redeeming exploits are those in association with another tiny

Mosquito fern (*Azolla caroliniana*) forms floating aquatic mats and turns red in full sun. Photo by Jack Goldfarb, courtesy of Design Pics.

organism, an unassuming microscopic blue-green bacterium called *Anabaena azollae*. Tiny filaments of this photosynthetic bacterium love to nestle into cavities formed by the overlapping leaves of *Azolla* ferns. Surrounded by this host, the bacteria carry out the amazing process of nitrogen fixation (taking nitrogen from the air and converting it into a form that plants can use as fertilizer). This phenomenon has been capitalized upon by rice farmers in Asia since at least the eleventh century. *Anabaena*-laden water fern is grown on the surface of rice paddies. When the paddies are drained to harvest the rice, the fern and its bacteria partner decompose on the soil and provide a rich source of nitrogen fertilizer for the next rice crop. As a bonus, the thick carpet of water fern makes it difficult for mosquito larvae to get to the surface for air—definitely an advantage for those working around still water. That's some kind of fern!

BLUE OIL FERN
Microsorum thailandicum

Did You Know?
Very few plants are this iridescently blue—a feature thought to be an adaptation to low light conditions, allowing the plant to harvest more of the light that is available.

PLANT TYPE: Herbaceous perennial
HEIGHT AND SPREAD: 3 in. × 8 in. (8 cm × 20 cm)

NOT ONLY DOES THIS BEAUTY not look like a fern, it doesn't even look like a real plant! Its fronds are strap-shaped (undivided, not ferny), thick-textured, and frankly—blue. Not just with a bluish cast like certain hostas, but shiny, iridescent, almost cobalt blue. The name for this plant in its native Thailand is *waew peek maeng thub* or "shiny as the wings of a scarab beetle." It is amazing that such a unique and gorgeous plant was only first described as new to science in 2001! Nature still holds mysteries and treasures undiscovered, just one of many reasons to treat her a little better.

Blue oil fern grows in moist, warm limestone hills, usually epiphytically (above the ground on another plant) or lithophytically (in rock crevices). How enchanting it would be to walk in that jungle and look around to see clumps of long, shiny blue leaves appearing to sprout directly from the sides of trees and rocks. And such a scene would be real, not hallucinogenically produced! Very few plants are this iridescently blue—a feature thought to be an adaptation to low light conditions, allowing the plant to harvest more of the light that is available.

Understandably, *Microsorum thailandicum* caused a minisensation among fern and terrarium enthusiasts when it first hit the market, fetching top dollar. Nei-

Blue oil fern (*Microsorum thailandicum*) has an eerie sheen on the shady forest floor. Photo by Richard H. Gross.

ther difficult nor easy to grow (give it what it likes consistently), this special fern is now more available to the general public. If you want to give it a try, a terrarium or greenhouse would be ideal. Of course you might be able to fool a few folks into thinking you own this plant by just buying a plastic strap-shaped plant and painting it with shiny blue nail polish. That wouldn't be a weird plant, though—it'd just be plain weird.

STAGHORN FERN
Platycerium

PLANT TYPE: Herbaceous perennial
HEIGHT AND SPREAD: 24 in. × 24 in. (61 cm × 61 cm)

LIKE GIANT, GREEN BARNACLES on the side of a piling, staghorn ferns cling to tree trunks, forming a rather odd scene to the first-time viewer. These ferns are huge epiphytes, and a large, mature clump can weigh up to several hundred pounds (kilograms). Besides sticking to the sides of trees, they are quite unique in form, having unfernlike (no lacy fern leaves here) and dimorphic (two distinct types) fronds. In some commonly grown species, one type of frond looks like silvery green, drooping deer or moose antlers complete with a fine, fuzzy coating. This appearance gives the plant its common and scientific names (*Platycerium* means "broad horn" in Latin). In different species, these forked fronds can also be compact and stiffly ascending, or can hang down with wobbly, twisted segments like a massive head of green tresses. Patches of brown spore-producing tissue form on the undersides of these fertile fronds.

The other type of frond starts out as a flat green circle of tissue growing over the surface of a substrate, be it tree trunk in the wild or clay pot in cultivation. These are the grasping, adhesive fronds, and they do

Giant staghorn fern (*Platycerium superbum*) has spore-producing hanging fronds and sterile upright fronds. Photo by M. Flagg, courtesy of the Australian National Botanic Gardens.

not stay green for long, quickly turning brown and papery. However, they are tough and long-lasting, and the buildup of these "dead" shield-like fronds is crucial to the health of the plant. They form a protective sponge, covering the actual roots of the fern, and collecting decaying matter that falls from above, like a giant wastebasket creating an aerial compost pile to provide nutrients and retain moisture.

Several *Platycerium* species are suitable for growing in bright light indoors or partial shade outdoors in summer, with the most commonly available being selections of *P. bifurcatum*. While these ferns can be grown in solid hanging baskets, a more "natural" method is to attach them to a large piece of redwood or cork bark, and they can often be purchased premounted. I like this system because you can hang your ferns neatly on the wall with little plaques announcing when you bagged each trophy staghorn.

Fern Allies—Alone and Frondless

Before there were true ferns, there were even more primitive land plants. We call these the fern allies, and they arose some 450 million years ago to colonize the land as the first vascular plants. The earliest lacked true leaves, or fronds, but still reproduced by spores. Even with their simple structures they dominated the earth, and some grew into trees. During the Coal Age, joined by the ferns, they reached their apogee, forming forests of giant seedless plants. Their glory lives on today mainly as widespread coal deposits. These plants are all but extinct, having given way on an ever-cooling and drying earth to the more adaptable true ferns, conifers, and flowering plants. We present here three of the last remaining groups of the most primitive plants on earth: scouring-rushes, clubmosses, and whisk ferns.

Rock tassel-fern or clubmoss (*Lycopodium squarrosum*) is a tropical epiphyte with forking branches. Photo by Richard H. Gross.

SCOURING-RUSH
Equisetum

PLANT TYPE: Herbaceous perennial
OTHER COMMON NAME: Horsetail
HEIGHT AND SPREAD: 6 in. × 36 in. (15 cm × 91 cm)

JEOPARDY CLUE: "This plant can clean your dishes in a pinch." Most of us would switch categories after that clue, but if the ghosts of the pioneers could ring in, they would be quick to answer, "What is scouring-rush?" Out on the Midwestern prairies in pioneer times, there were no detergents, no Brillo pads, and certainly no Teflon. Pioneers cooked in cast-iron pots over open fires—a recipe for caked-on grease. Those were some resourceful folks, though, and they discovered a plant growing abundantly near water with a surface so tough it could actually be used to scrub cookware. The stiff, rodlike stems of scouring-rush have sand grains (silica) embedded in their textured ribs, and these grains act like a crude steel wool. The pioneers gave the plant its common name. Today it is often called horsetail, a translation of its Latin name, *Equisetum*.

Equisetums are the surviving members of an ancient group of fern relatives, lacking true leaves, bearing only a ring of gray pointed scales at each joint on the stem. The green stems carry on photosynthesis, and the plant reproduces by spores produced in a conelike structure at the tip of each reproductive stem. Only about fifteen species of these living fossils are left on earth. The largest, *Equisetum giganteum*,

---

GROWING TIPS

DIFFICULTY RATING: **1**
INDOORS VS. OUTDOORS: Indoors or outdoors
LIGHT: Full sun to bright shade
HARDINESS: Zone 4
MOISTURE: Keep wet to moist
GROWING MEDIUM: Sandy clay outdoors or good, well-draining potting mix
WARNING: Spreads by underground rhizomes and is difficult to eradicate. No common chemicals kill it. Do not plant in uncontained outdoor situations

---

Did You Know?
The stiff, rodlike stems of scouring-rush have sand grains (silica) embedded in their textured ribs, and these grains act like a crude steel wool.

Scouring-rush (*Equisetum hyemale*) has jointed, ribbed stems and a sheath of scalelike leaves at each joint node. Photo by Richard H. Gross.

grows in Central and South America and may reach 20 ft. (6 m) tall. Impressive, but still just a shadow of the giant *Equisetum*-like *Calamites* trees that formed vast forests during the Coal Age, 300 million years ago.

The common scouring-rush, *Equisetum hyemale*, produces stiff, pencil-sized, jointed stems to 4 ft. (1.2 m) tall. No one knows why it deposits silica in its stems. It might strengthen them and prevent herbivory. The field horsetail, *E. arvense*, bears wispy branches at each stem node, creating the look of a tail of coarse hairs. Both species are relatively common, growing in locally moist areas in large patches. You can enjoy them by planting them in their own pots to contain their aggressive growth. Every now and then sacrifice a few stems to scrub off the barbecue, to remind you of a time when people looked directly to plants to make their daily lives a little better.

CLUBMOSS
Lycopodium

PLANT TYPE: Herbaceous perennial
OTHER COMMON NAME: Ground pine
HEIGHT AND SPREAD: 12 in. × 24 in. (30 cm × 61 cm.)

THE CHINESE INVENTED FIREWORKS 2000 years ago. Someone put gunpowder in a bamboo tube, and it exploded when burned. But clubmosses (*Lycopodium*) have had the capacity to do that for millions of years. They have "pyrotechnic" spores, filled with volatile plant oils that are highly flammable (we're not sure why, but perhaps to enhance longevity). In the early days of photography, illumination was supplied by flash powder spread on a tray, which when ignited would burst into a fireball of brilliant light. This powder was the spores of clubmosses, laboriously collected by hand from countless plants during late summer. Early photographs were warmly gratifying, as these icons of loved ones could be had more cheaply than oil paintings.

The spores of *Lycopodium* species have proven useful in their nonignited state as well. For centuries these spore powders have been used medicinally for treating skin problems and dusting pills to prevent sticking. In the early twentieth century they were used to coat the inside of latex gloves and condoms, acting like microscopic ball bearings to form a nongreasy lubricant.

Clubmosses, also known as ground pines, are related to neither mosses nor pines. They are primitive

GROWING TIPS

DIFFICULTY RATING: **2**
INDOORS VS. OUTDOORS: Tropical species in a humid greenhouse or plant room. Temperate species can be found growing outdoors in the woods but are unobtainable commercially and not transplantable
LIGHT: Bright shade
HARDINESS: Tropical species not hardy
MOISTURE: Keep moist
GROWING MEDIUM: Epiphyte mix of 50% potting mix, 50% perlite

Did You Know?
In the early days of photography, illumination was supplied by flash powder spread on a tray, which when ignited would burst into a fireball of brilliant light. This powder was the spores of clubmosses, laboriously collected by hand from countless plants during late summer.

An American clubmoss (*Lycopodium clavatum*) cone produces abundant pyrotechnic spores that burst into flame when ignited. Photo by Larry Mellichamp.

vascular plants that reproduce by spores only and whose stems bear tiny, simple leaves. Most temperate species are low-growing, creeping plants, but tropical species can be quite substantial, hanging from trees limbs in large masses, stems dangling and forking in twos as they grow. These are the types that earn the label of "weird," even without knowledge of their flammable spores. Those abundant spores are produced in sacs at the base of leaves or in tightly formed, erect, conelike structures.

Fewer than 400 species still exist on earth, and they represent a bygone era of plant life. These ancient plants, and their frondless relatives the scouring-rushes, were part of the Coal Age forests that formed our coal deposits some 300 million years ago. Even in death, they serve to strike fire and continue to warm the hearts of mankind.

WHISK FERN
Psilotum nudum

PLANT TYPE: Herbaceous perennial
HEIGHT AND SPREAD: 12 in. × 12 in. (30 cm × 30 cm)

IT'S GOOD TO START OUT SIMPLE, particularly if you are trying something for the first time. Like living on land, for instance. Believe it or not, you can find alive on earth today a plant that is remarkably similar to the fossilized specimens that scientists have called the first (true) land plants from some 450 million years ago. And simple it is—the whisk fern (*Psilotum nudum*) lacks true roots, leaves, flowers, fruits, and seeds. It has only dichotomously forking stems (a very rare trait) and produces spores in small yellowish sacs along its stem. This lends an odd, minimalist appearance to the plant.

While mostly existing in tropical forests all around the world, whisk ferns are also found in North America from the Florida Everglades to as far north as the Great Dismal Swamp in southeastern Virginia. Their spores are lightweight and easily float about in air currents, although once settled they take many years to develop into new plants. This hasn't stopped them from becoming real pests in greenhouses worldwide. They don't seem to be picky about their company, growing in pots of dry-land succulents, epiphytic orchids, and shrubs alike. You can also spot them growing out between the mortar cracks in old brick walls in Charleston, South Carolina.

Once established, whisk ferns are long-lived, undemanding, and make a most unusual specimen—a true living fossil. Having been around for so long (tens of millions of years) means that these plants have outlived their close relatives, not to mention their diseases. And without the leafy stems of the more advanced ferns, they are therefore without fronds. They need us to appreciate them, for otherwise they are simply alone and frondless in this modern world. How long will they continue to exist? They have already lived through all the changes the earth has endured. Perhaps we need to learn their secret.

Whisk fern (*Psilotum nudum*), the world's simplest vascular plant, produces rootless, leafless, dichotomously forking stems. Photo by Larry Mellichamp.

3

FLAMBOYANT FLOWERS AND BRIC-A-BRACT

Natural Sexual Enhancement

AT A BIG PARTY, there is always one who stands out from the crowd and just begs to be looked at. If she is going to wear a hat, she wears a *big* hat. But although the eyes of the crowd can't help but move in her direction, strike up a conversation with the flamboyant dresser, and she may not always have a lot to say behind all the flash. It's all about advertising and attraction. Think about billboards along the highway. Some are simple—"Eat at Joe's"—while others proclaim loudly in neon, "Best barbecue in the state!" The claim is usually untrue, but who can resist the suggestion?

Everybody knows that flowers produce nectar, and normally have showy petals to advertise that fact. Some plants are not satisfied with that and seem to want to go way beyond the norm into the realm of hype and grandiose displays of floral fancy. But, of course, there is a lot at stake (that is, sexual reproduction and the viability of that species' future!). So if a plant needs that extra boost to attract a pollinator, it might as well pull out all the stops. Why stop at petals for attraction, when you have sepals, bracts, and even stamens right there that could be pushed into service? Dress 'em up and put 'em on the curb! After all, plants have the "plasticity" to modify or exaggerate organs when needed. For instance, if they need to keep insects away from their buds, they can make sticky

---

Hanging lobster claw (*Heliconia rostrata*) has colorful bracts on a long, pendulous inflorescence. Photo by Richard H. Gross.

hairs on the stem. Or if they need to reflect light they can cover a leaf in a mat of white hairs.

Bracts are simply leaves associated with flowers. They can be large or small, and they're one of the best tools in the plant's arsenal for enhancing its floral display. They can be greatly exaggerated and gaudily colored, providing an all-out flag-waving, banner-busting, broadside-beaming shout-out for attention. But creating such large structures can have its drawbacks. It costs more in resources and can make the flowering branches heavier when it rains, leading to breakage. It must be worth the risk. The investment could pay off in a crowded habitat or when pollinators are scarce. Maybe the plant produces low-quality nectar and has evolved to advertise more loudly in order to appear to have the better product. Perhaps it just can't make big flowers because of the nature of its family traits, as with euphorbias or the many plants in the arum family (Araceae). A well-known example is the flowering dogwood tree of eastern North America. Its four white "petals" are actually bracts. The real flowers are tiny yellowish things aggregated in the middle. Without the white bracts, we (and, I assume, pollinators) wouldn't give the flowers a second glance. So can you blame these plants for puttin' on the Ritz?

TITAN ARUM
Amorphophallus titanum

PLANT TYPE: Herbaceous perennial
OTHER COMMON NAME: Corpse flower
HEIGHT AND SPREAD: 10 ft. × 8 ft. (3 m × 2.4 m)

BOTANICAL GARDENS, AS IMPORTANT and wonderful as they are, don't get many rock star moments in the media. They have no pandas, polar bears, or IMAX reenactments of battling dinosaurs. A growing number of them do, however, have one unpredictable ace up their sleeves. It goes by the name of *Amorphophallus titanum*, or titan arum, and it may just be the world's most famous plant. Want to see lines of visitors forming outside your garden? Bloom one of these giants.

Titan arum was discovered in 1878 in Sumatra by the Florentine botanist Odoardo Beccari and introduced to the Western world at Kew Gardens in London, where it flowered for the first time in 1889. From that moment onward, whenever and wherever one has bloomed in "captivity," it becomes all the rage—for a few days. Is the adoring public fickle? No. The sheer size and hair-raising stink of this outlandish "flower" could draw crowds for longer, I think. It is the flower itself that is fickle. Once open, it lasts a mere three to four days, then collapses like a behemoth into a limp, vegetative heap, ready for the compost pile.

Its enormous size, odd shape and color, rareness, fleeting nature, and the raw appeal of its stench have

GROWING TIPS

DIFFICULTY RATING: **3**
INDOORS VS. OUTDOORS: Indoors
LIGHT: Strong but filtered light. No direct sun
HARDINESS: Not hardy
MOISTURE: Keep constantly moist but not sopping wet, drier when dormant
GROWING MEDIUM: Good, well-draining potting mix
NOTES: Has an irregular cycle of growing followed by dormancy. May take seven to nine years or more from a small corm to flower. Does not like to go below 78°F (26°C) or above 90°F (32°C) but may briefly tolerate temperatures outside this range

Did You Know?
Amorphophallus *means "penis without shape," and some might agree that this is an apt description of the giant spadix emerging from the middle of the spathe.*

Newly opened titan arum (*Amorphophallus titanum*) inflorescence with temperature probe in the spadix extension. This
plant can raise its own temperature by about 15 degrees
above ambient to volatilize is powerful attracting odor.
Photo by Larry Mellichamp.

made a star out of this plant, which boasts the world's largest unbranched
inflorescence. What appears as a single, 4–9 ft. (1.2–2.7 m) tall flower, is
actually made up of a thick, fleshy axis (spadix) embedded with hundreds
of separate male and female flowers, surrounded by a gigantic, deep
purple-red, cloaklike bract that looks like a 4–6 ft. (1.2–1.8 m) wide petal.
The technical name for this unfolding bract is "spathe." *Amorphophallus*
means "penis without shape," and some might agree that this is an apt
description of the giant spadix emerging from the middle of the spathe.

The upper three-quarters of that spadix actually generates heat in order to emanate an odor that is most often described as that of rotting
flesh and that can carry on the wind for a quarter-mile. In the plant's
natural habitat this odor serves to call in the pollinating troops of carrion beetles and flesh flies, who must imagine, to their delight, that a jungle
beast has met an unfortunate end. Looking for flesh to eat and lay eggs in,
these visitors crawl down into the spathe and gather at the bottom, then
spend the night wondering where the party is. Covered with pollen the
next morning, they escape from this venue, only to be lured by the promise of another stinky hot spot several blocks away, where female-recep-
tive flowers wait down inside another cloak of purple. In this large-scale
deception, the somewhat widely scattered individual plants achieve the
cross-pollination required to produce seeds.

It takes eight to twelve years before the plant blooms from seed. What
happens in the meantime? The vegetative plant grows from a very large
underground corm (tuberlike structure) and consists of a single, but
branched, 8–15 ft. (2.4–4.6 m) tall, upright leaf. This leaf looks more like
a strange, green, speckled and spotted tree or perforated giant beach um-
brella. The plant goes through cycles of leaf growth and dormancy, during which time (probably the dry season in its native jungles) there is no

aboveground evidence of the plant. At each cycle, the leaf and tuber grow larger. Then, at some unknown point, the next growth cycle produces an inflorescence instead of a leaf. It takes several weeks from the time when you can tell the new sprout will be an inflorescence until the spathe opens. The days leading up to the opening are filled with anticipation and excitement. It is much like the countdown to the birth of baby—every day you measure the height and girth, remark on how much it has grown, and speculate on exactly when the blessed event will take place. Then, bam! It happens at five o'clock in the morning when you least expect it. You go running breathlessly to see it, and—wow—it is amazing beyond belief. The colors are vivid, the smell is unbearable, and to be in its presence is simply awesome.

Can you grow a titan arum at home? Yes, you can. People have done it in a backyard greenhouse. But do you want to? They are expensive as initial corms, though increasingly available. They take years of cycling between growing and going dormant to reach their optimal size, during which time they like it quite warm, up to 90°F (32°C), and humid in summer, and constantly well above 70°F (21°C) in winter. They like strong but filtered light—no direct, hot sun. They need a big pot, good potting soil, and lots of water and fertilizer, but shouldn't be kept too wet when dormant. Our cycle at the UNC Charlotte Botanical Gardens was one and a half years of leaf growth, then half a year of dormancy—for six years. It finally bloomed on 1 July 2007. We had failed once before when it got too cold one year—40°F (4°C) for a week—but we tried again. I say, go for it!

PIGTAIL PLANT
Anthurium scherzerianum

PLANT TYPE: Herbaceous perennial
OTHER COMMON NAME: Flamingo flower
HEIGHT AND SPREAD: 12 in. × 12 in. (30 cm × 30 cm)

IF YOU THINK WE HAVE ALL BECOME too sophisticated and it takes a lot to pique our interests, hang around a pigtail plant at a botanical garden for a few hours. Both young and old are amused by the orange

GROWING TIPS

DIFFICULTY RATING: **1**
INDOORS VS. OUTDOORS: Indoors but may be grown outdoors in summer
LIGHT: Moderately bright light indoors. Partial shade outdoors
HARDINESS: Not hardy
MOISTURE: Keep evenly watered. Likes higher than average humidity
GROWING MEDIUM: Well-draining potting mix with extra perlite. Fertilize twice monthly
NOTES: Avoid dividing until pot-bound

Pigtail plant (*Anthurium scherzerianum*) has a curious curled spadix of embedded miniscule flowers, but the spathe is the big show. Photo by Paula Gross.

curlicues of the "flower" and are especially delighted when they hear its common name. Occasionally I get a blank stare when I point out that it looks like pig's tail. (I guess farmyard animals don't get the media exposure they used to.) But luckily, thanks to Babe, pig of movie fame, most folks do still know what a real pig's tail looks like.

Botanically, that pig's tail is part of a special inflorescence type called a spathe and spadix, a key characteristic of the huge plant family, Araceae, to which *Anthurium scherzerianum* belongs. The spathe is a bract, in this case particularly large and brilliantly red, that surrounds or hangs below the usually elongated spadix, in this case what looks like a pig's tail. The spadix bears the actual flowers, but you would never know it from looking. They are miniscule, consisting only of anthers or pistils embedded in the spadix.

The other common name for our plant, flamingo flower, is shared by another species, *Anthurium andraeanum*, which has large, shiny red, pink, or white spathes and straight yellow spadixes. I never really understood the common name of flamingo flower until I saw a particular botanical illustration of it. Previously I thought it must simply refer to the salmon pink color of the spathes of some *A. andraeanum* plants. In the illustration I saw the spadix arched forward in just such a way as to mimic the neck of a flamingo, while the spathe hung down, mimicking the size and shape of a flamingo's body. Wow, how cool! I can't wait to find one in the greenhouse that looks like that—a new "animal" to show visitors. And thanks to the kitschy popularity of plastic yard art, they will recognize just what they are looking at.

DUTCHMAN'S-PIPE

Aristolochia

PLANT TYPE: Woody perennial vine
OTHER COMMON NAMES: Pipevine, birthwort
HEIGHT AND SPREAD: 3–20 ft. × 1 ft. (0.9–6 m × 0.3 m)

AROUND 500 SPECIES of *Aristolochia* (Aristolochiaceae) are recognized worldwide, temperate and tropical, and a great many of them bear the strangest flowers. Those from the tropics often have the largest flowers, and A. *grandiflora* from Central America has a flower that is much "too big for its plants." The flowers can be over 1 ft. (30 cm) long and 8 in. (20 cm) wide and seem out of proportion to the heart-shaped leaves and slender vine. A typical reaction is to stare in amazement and then approach with caution, as if the flowers might have a slightly ominous life of their own.

As is always the case, the life of the showy flower is all about attraction and securing pollination. Aristolochias seem to have hit upon a combination of unique shape, coloration, and odor that works, as all species share certain floral traits and tend to attract flies as pollinators. The flowers have one other peculiarity: they are formed of sepals only, there are no petals. The flowers of *Aristolochia grandiflora*, viewed from the front, look like stiff, yet also floppy, washcloths—a little rough in texture and patterned like burgundy batik, with a long tail hanging down from the lower margin. Somewhat heart-shaped from the front, the flower from the side reveals a curving floral tube, shaped like a stretched S-curve. This side-view

GROWING TIPS

DIFFICULTY RATING: **1**
INDOORS VS. OUTDOORS: Indoors or outdoors
LIGHT: Mostly sun for tropical species. Semishade for temperate species
HARDINESS: Hardy species to zone 6. Tropicals not below 45°F (7°C)
MOISTURE: Keep moist
GROWING MEDIUM: Hardy species in well-draining garden soil. Tropicals in well-draining potting mix in large container. Fertilize twice monthly

Did You Know?

Luckily the scent is not too overpowering for human admirers, but flies sure go for it. Thinking they have found rotting meat in which to lay their eggs, they enter the flower and travel down the curving floral tube.

A large Dutchman's-pipe (*Aristolochia grandiflora*) flower
produces a long tail and fly-attracting odor—for one day.
Photo by Paula Gross.

shape, characteristic of many of the species' flowers, apparently looks
enough like a Dutch smoking pipe to give the genus its common name.
Other species' unopened flowers reminded medieval herbalists of the
birth canal or a small fetus, earning them the name of birthwort and a
place in the medicine chest (thanks to the Doctrine of Signatures—see
chapter 4). When the flower opens, it exudes a fetid odor. Luckily the
scent is not too overpowering for human admirers, but flies sure go for it.
Thinking they have found rotting meat in which to lay their eggs, they
enter the flower and travel down the curving floral tube. In a freshly
opened flower, the tube is coated inside with many long hairs that prevent the flies from leaving. The next morning they are showered in pollen, and the hairs wither, allowing the "guests" to fly out. Won't they ever
learn about this bamboozle? The flowers hope not.

Some insects, namely certain swallowtail butterflies, use the Dutchman's-pipe to their advantage, not vice versa. Their larvae feed almost
exclusively on *Aristolochia* foliage from which they get a nasty chemical
they sequester in their bodies to make them distasteful to birds. Whether
feeding caterpillars, trapping flies for an obligatory evening inside their
flowers, or mesmerizing humans, aristolochias simply will not be
ignored. Put that in your pipe and smoke it!

SEA HOLLY
Eryngium

PLANT TYPE: Herbaceous perennial
HEIGHT AND SPREAD: 24 in. × 12 in. (61 cm × 30 cm),
  taller for some types

THE FLOWERING STALKS OF SEA HOLLY are not delicate beauties but are rather striking and architectural. To me they look like they've been cut out of aluminum and lightly spray-painted a pale, somewhat unnatural-looking blue. Their foliage is stiff, sharp-edged, and often gray. I have seen metal "flower arrangements" at craft fairs that look amazingly similar to this, and I could see sea holly as the model for sixties-era space-age flowers. But sea holly and other eryngiums come by their appearance naturally.

More than 200 species of *Eryngium*, herbaceous members of the carrot family (Apiaceae), are scattered across the globe—a mix of annuals, biennials, and perennials. While not all are showy and dramatic, they do share a similar pattern in their floral structure. What looks like a cut-metal, spiny set of petals surrounding a round or cone-shaped structure like a pincushion full of pins is actually an inflorescence. Each little "pin" or unit in the center portion is a tiny, fully formed, functioning flower complete with male and female parts. They are arranged tightly together, radiating from a central point into what botanists call

Sea holly (*Eryngium*) has striking "metallic" bracts surrounding a head of tiny flowers. Photo by Larry Mellichamp.

an umbel. Below this umbel are the large, showy bracts—what we might have interpreted as petals. The bracts combined with the group of tiny flowers serve as a much bigger billboard to roaming insects than the true individual flowers themselves. The silvery gray of foliage and bracts may not be particularly enticing to daytime pollinators, but the gray coating may serve as a reflecting sunscreen, as eryngiums tend to be full-sun, coastal or more often grassland plants. The eerie blueness of some species' flowers and bracts appeals to bees and gardeners alike.

Several species and hybrids are useful garden plants, including *Eryngium planum*, *E. bourgatii*, and *E. giganteum*. There is even an eastern United States native, *E. yuccifolium*, or rattlesnake master, that is becoming popular with gardeners for its unique appearance. Grow sea hollies to bring a little steely drama to your garden. They make great cut flowers, especially if you are into that postmodern aluminum look.

Gloriosa lily (*Gloriosa superba*), with its flamboyant flowers, is a vine that climbs by means of leaf-tip tendrils. Photo by Paula Gross.

GLORIOSA LILY
Gloriosa superba

PLANT TYPE: Herbaceous perennial
HEIGHT AND SPREAD: 36 in. × 12 in. (91 cm × 30 cm)

THIS AFRICAN LILY STANDS OUT from the crowd of its enormous plant family, the Liliaceae. Not only does it have flowers that look like flaming fireballs hurtling through space, but these flowers are born on $3\frac{1}{2}$ ft. (1 m) vines that climb by a botanically rare type of tendril: the tips of the elongated leaves themselves coil and grasp supports, like hands grabbing and climbing a ladder. Most vines climb by twining or by tendrils produced by modified stems or whole leaves, not just the tips.

Gloriosa lilies have wavy-edged sepals and petals that look alike (tepals), and come in fiery colors of yellow, orange, and red. To have the sepals colored like the petals is unusual and takes the place of developing colorful bracts. These tepals are strongly reflexed (swept backward), and the stamens and stigmas stick far out front, designed for a pollinator with a broad wingspan and long proboscis—a butterfly. Curiously, in gloriosa lily the style comes off the top of the ovary and quirks to one side at an abrupt angle, as if Mother Nature decided at the last minute that the style was not positioned correctly and put a joint there. This is yet another odd feature that makes this lily just a little off-kilter and therefore more lovable in our eyes.

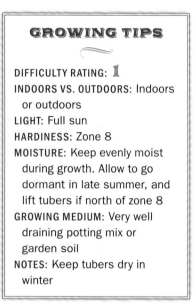

GROWING TIPS

DIFFICULTY RATING: **1**
INDOORS VS. OUTDOORS: Indoors or outdoors
LIGHT: Full sun
HARDINESS: Zone 8
MOISTURE: Keep evenly moist during growth. Allow to go dormant in late summer, and lift tubers if north of zone 8
GROWING MEDIUM: Very well draining potting mix or garden soil
NOTES: Keep tubers dry in winter

Did You Know?
Not only does it have flowers that look like flaming fireballs hurtling through space, but these flowers are born on 6 ft. (1.8 m) vines that climb by a botanically rare type of tendril: the tips of the elongated leaves themselves coil and grasp supports, like hands grabbing and climbing a ladder.

LOBSTER CLAW
Heliconia

PLANT TYPE: Herbaceous perennial
HEIGHT AND SPREAD: 36 in. × 36 in. (91 cm × 91 cm)

Rᴏᴜɴᴅɪɴɢ ᴛʜᴇ ᴄᴏʀɴᴇʀ in a conservatory and encountering the hanging inflorescences of *Heliconia rostrata* is a memorable experience (see page 84). After the initial exhilaration over their size and bright red-orange-yellow coloration, the brain starts working to categorize their form. The individual floral units (actually bracts) are proportioned just like the claws of a lobster. Stacked on top of each other, they look like a warning to wayward lobsters: "Your kind is not welcome here!" Or if you are feeling hungry, the pendant bracts may look like some avant-garde cook has carefully strung individual claws, alternating left and right to form a sort of edible mobile. Now that's presentation! The upright lobster claw, *H. stricta* (Heliconiaceae), is much smaller and more manageable by comparison.

Mother Nature may have a sense of humor, but as botanists we see these unique floral structures as perfect adaptations for hummingbird pollination. Each "claw" is actually a colorful, long-lasting, waxy bract. Inside the bracts, and poking out when mature, are the true flowers, which are tubular and often pale or

GROWING TIPS

DIFFICULTY RATING: **1**
INDOORS VS. OUTDOORS: Indoors but may be grown outdoors in summer
LIGHT: Bright light indoors. Partial sun outdoors
HARDINESS: Not hardy
MOISTURE: Keep well watered but well draining
GROWING MEDIUM: Well-draining potting mix. Fertilize twice monthly during growing season
NOTES: Divide periodically. Do not allow to go below 50°F (10°C)

Upright lobster claw (*Heliconia stricta*) has strikingly colored bracts surrounding dull-colored greenish flowers. Photo by Paula Gross.

greenish. Hummingbirds are attracted to the red bracts but soon discover the delightfully nectar-filled flowers, designed to deposit pollen on their heads or chins as they probe with their tongues.

Warning to collectors: There are more than 150 species of *Heliconia*. Not all look like lobster claws, but typically they are bold in form and strikingly colored. While most make very large specimens, one of the smaller types can lend a distinctly tropical air to the sunroom or back deck. Start out with a "dwarf" *H. stricta* 'Sharonii' hybrid, as these are happy in a large pot and grow only 4–6 ft. (1.2–1.8 m) tall. The South American *H. rostrata*, with its hanging lobster claws, can exceed 6 ft. and needs room to grow—a bit impractical (but not impossible) for home growers. Unless you have a problem with marauding lobsters, it might be best to appreciate *H. rostrata* at your local botanical garden conservatory.

HARDY BANANA
Musa

PLANT TYPE: Herbaceous perennial
HEIGHT AND SPREAD: Variable, to 12 ft. × 12 ft. (3.7 m
× 3.7 m)

THE COMMON BANANA TREE bestows an instant tropical look to plantings in temperate climates. Just add a piña colada and a Jimmy Buffett CD to those gigantic leaves swaying in the breeze, and the backyard swimming pool becomes an instant vacation. But flowers? Do banana trees even make flowers? Of course they do, for no flowers would mean no bananas. It's just that in the temperate zone, typical tropical banana species freeze to mush before getting old enough to make them—they require eighteen months of nonfreezing temperatures to mature the edible banana fruits we are used to buying from the grocery store. Grow one of the hardy banana species, though, and you will get more than bold leaves (although you won't get edible bananas, darn). You may be mighty surprised when late in summer from the top of the majestic crown of banana leaves springs an ever-unfolding series of colored bracts from a swollen podlike growth. Look closer at this erupting inflorescence and you can find neat rows of true flowers cupped by the large, leathery bracts. As long as temperatures remain above freezing, you will be treated to new flowers every few days as the inflorescence grows longer and longer!

GROWING TIPS

DIFFICULTY RATING: **1**
INDOORS VS. OUTDOORS: Indoors or outdoors
LIGHT: Full sun
HARDINESS: Variable, zones 7–10
MOISTURE: Provide lots of water. Not at all drought tolerant
GROWING MEDIUM: Good, well-draining potting mix or garden soil
NOTES: Larger edible bananas require eighteen months of nonfreezing temperatures to produce mature fruit

Pink banana (*Musa velutina*) produces showy pink bracts and pink fruits all summer. Photo by Larry Mellichamp.

The Japanese fiber banana (*Musa basjoo*, in the banana family, Musaceae) is already known to be hardy to below 0°F (−18°C) because it comes from the colder regions of China and has been introduced into Japan. (Perhaps we should start calling it the hardy Chinese banana.) It is a striking, large plant, to 15 ft. (4.6 m) tall, but it will die down and grow back up quickly after each winter without being brought indoors. In a mild winter it may actually remain standing, leaf out even earlier, and get even taller. In the South, even on the coldest nights the temperature usually dips down and then back up, and the ground does not freeze for long periods. With protection and the use of microclimates, we can grow things we should not be able to. We have certainly pushed the envelope since the 1990s by growing bananas, palms, gingers, elephant's ears, and other plants we "knew" were not hardy, through the coaxing of plantspeople such as the late J. C. Raulston, nurseryman Tony Avent, and members of the Southeastern Palm Society.

As we continue to try new plants, we have discovered that we especially like the hardy pink banana, *Musa velutina*, because the entire plant is smaller, to just 10 ft. (3 m), including its leaves, but it still has the bold banana look. Furthermore it produces showy yellow flowers all summer in a colorful pink-bracted inflorescence held at eye level. The pink fruits open when ripe in late summer and do smell like bananas but are too seedy to eat (although we know you can't resist trying them!). If you want to grow your own eating bananas, you will need a heated greenhouse.

Finally, a warning: Avoid getting the sap of the banana plant (affectionately called "banana blood") on your body or clothing. It is like invisible ink—clear when wet, darkly staining as it dries—only it will stain cloth permanently. On the other hand, experiencing this sort of rite of passage tells other gardeners that you are a botanical adventurer and will grow bananas where no bananas have gone before.

PASSIONFLOWER
Passiflora

Did You Know?

*In the elaborate flowers of a
vine, early Jesuit priests saw
the entire story of the crucifixion, or passion, of Christ, and
named it passionflower.*

PLANT TYPE: Herbaceous or woody perennial vine
HEIGHT AND SPREAD: 72–144 in. × 6 in. (183–366 cm
  × 15 cm)

WHAT MAKES PASSIONFLOWERS WEIRD? These plants are both striking and beautiful, like the exotic women in the burlesque tents of old, not so much weird as mesmerizing in their elaborate (albeit somewhat revealing) costumes. Probably the strangest feature of the passionflower is its frilly corona ("crown"), the ring of tissue surrounding the ovary of the flower. The multitude of threadlike segments radiate out like rays of the sun or angelic halos, bearing rings of different colors. We really do not know the purpose of

Passionflower (*Passiflora*) tells quite a story by means of its uniquely structured flower. Photo by Larry Mellichamp.

these floral accoutrements, but we assume they must be helpful in pollination. We might guess that the passionflower got its name from these
"flames" of passion so prominent in the flower, but not so. *Passiflora* has
another story to tell.

I can imagine that the Jesuit priests exploring (converting the natives)
in South American jungles in the sixteenth century were looking for
signs of encouragement—divine symbols to assure themselves that their
God was present. So, in the elaborate flowers of a vine, they saw the entire story of the crucifixion, or passion, of Christ, and named it passionflower. Explanations differ somewhat but usually go something like this:
the tendrils of the plant represent the whips used on Christ, the radial
filaments are the crown of thorns, the central floral column is the pillar
of scourging, the three stigmas are the three nails, the five lower anthers
are the five wounds, and the ten sepals and petals are the ten *faithful* apos-
tles. Wow! I wonder what other religions might see in this flower.

Hundreds of species of *Passiflora* (Passifloraceae) are recognized,
mainly in the Americas and Asia, and they have much more than religious symbolism and beauty to offer. Many species are host to the larvae
of longwing butterflies, as well as providing nectar to bees and hummingbirds. A handful are cultivated for their edible "passion fruits,"
which are especially popular in juice blends. Finally, the leaves and roots
of the North American native P. *incarnata*, commonly called maypop because the inflated fruits "pop" when smashed, are used in herbal medicines and teas as a calming, nonaddictive sedative—an ironic use for a
plant with "passion" in its name. It may be that sometimes beauty really
does tame the beast. *Passiflora incarnata* is hardy to zone 5 and undemanding in its soil requirements.

BRAZILIAN CANDLES
Pavonia ×gledhillii

PLANT TYPE: Woody perennial shrub
HEIGHT AND SPREAD: 96 in. × 12 in. (240 cm × 30 cm)

PAVONIA ×GLEDHILLII (syn. *P. intermedia, P. multiflora, Triplochlamys multiflora*) is a Brazilian member of the mallow family (Malvaceae) with too many Latin names and, until recently, no common name. This has been unfortunate public relations, for the plant itself is bold, easy to grow, and almost constantly in bloom. Its screaming pink "flowers" are evocative of little bursts of fireworks, complete with a trail of blue sparks in their centers (blue pollen clinging to anthers).

Brazilian candles, as it is now known, can grow to 8 ft. (2.4 m) tall if left unpruned and is covered with flowers consisting of rings of reddish pink bracts surrounding the dusky purple petals that "cloak" the column of many stamens. The sepals and petals by themselves would be rather inconspicuous. It is the many strappy pink bracts that attract attention and guide the visitor (human or insect) to notice that unusual blue pollen decorating the stamens. And attract they do. Visitors at the UNC Charlotte Botanical Gardens often ask for the common name of this plant. I usually enjoy answering questions, but until recently I always dreaded this one, because for years this plant had no well-recorded common name. I was relieved to finally see a consistent name applied to it. I think

Brazilian candles (Pavonia ×gledhillii) has hot pink bracts surrounding a some-what quiet flower. Photo by Larry Mellichamp.

the flowers are much more dramatic than simple candles, but no mind. I am happy to see *Pavonia ×gledhillii* lighting up the greenhouse and be able to tell visitors, without a long explanation, "Brazilian candles. That's its name."

SPIDER'S TRESSES
Strophanthus preussii

PLANT TYPE: Woody perennial shrub
OTHER COMMON NAMES: Tasselvine, poison arrow vine
HEIGHT AND SPREAD: 72 in. × 24 in. (183 cm × 61 cm)

THIS BABY'S GOT PETALS that just won't quit! Literally. The petals of this vinelike, tropical African shrub narrow down into threadlike segments that dangle and intertwine with each other for lengths of up to 8 in. (20 cm). A blooming *Strophanthus preussii* has impact, covered with white or pale yellow and burgundy flowers, all elegantly draping their "tassels." So at least the common name of tasselvine makes sense.

"Spider's tresses" seems somehow appropriate, too, until you think about it. What spider has tresses? Tresses are long locks of hair. I can see that, but I am still not sure where the spiders come in. Another common name, poison arrow vine, is the most threatening and exciting. Do the flowers look like arrows? No, so it must mean that the plant bears poison of the sort that is powerful enough to drop a monkey dead out of a tree in one hit! Well, almost. It turns out that ouabain, the powerful and famous dart poison, does in fact come from *Strophanthus* plants—different species, though (*S. gratus*, *S. hispidus*, *S. kombe*). Strophan-

GROWING TIPS

DIFFICULTY RATING: **1**
INDOORS VS. OUTDOORS: Indoors but may be grown outdoors in summer
LIGHT: Very bright light to full morning sun
HARDINESS: Not hardy
MOISTURE: Keep evenly moist to dryish
GROWING MEDIUM: Well-draining potting mix

Did You Know?
In the past it was used in carefully measured doses as a treatment for congestive heart failure. It was also used, and still is used, by West African tribes in unmeasured doses, liberally smeared on arrow tips used to kill animals.

Spider's tresses (*Strophanthus preussii*) has long, twisted petal tips, but for what purpose? Photo by Larry Mellichamp.

thus preussii seeds do contain much smaller amounts of very similar cardiac poisons.

Ouabain represents a great example of how the difference between a plant medicine and a plant poison is simply a matter of dosage. In the past it was used in carefully measured doses as a treatment for congestive heart failure. It was also used, and still is used, by West African tribes in unmeasured doses, liberally smeared on arrow tips used to kill animals.

Described as vinelike, *Strophanthus preussii* has neither tendrils nor twining stems. It is more of a sprawler, relying on other plants or, in cultivation, a frame or trellis to keep it upright. Whether you enjoy it for its dark red tassels or the fact that it contains arrow poisons (in very small amounts), *S. preussii* is a great addition to an arsenal of weird plants, spider's tresses and all.

BAT PLANT

Tacca

PLANT TYPE: Herbaceous perennial
HEIGHT AND SPREAD: 36 in. × 36 in. (91 cm × 91 cm)

BAT PLANT—THAT SOUNDS INTRIGUING. Do the flowers look like bats? No. It must be pollinated by bats. No. Oh, maybe bats live or roost on the plant? No. If your interest is waning, be patient—the common name will be explained. To see *Tacca chantrieri* or *T. integrifolia* in flower is enough to capture your attention without ever having heard their gothic common name.

The floral display of these tropical Southeast Asian plants is nothing if not extravagant. Emerging from among the clumps of broad, shiny leaves are tall floral stalks ending in two large, winglike bracts held above or behind a cluster of elegantly arranged, fleshy black flowers. From among the flowers emerge long, whiskerlike bracts that can reach lengths of 2 ft. (61 cm) or more in the white bat plant (*Tacca integrifolia*). Some folks think the wings of the black bat plant (*T. chantrieri*) earn its common name, but the overall inflorescence doesn't really look like a bat. Such an investment in floral accoutrements must mean that this plant is after an elusive pollinator. Yet scientists can't figure this out—the plants appear to be self-pollinating! The fleshy flowers grow into pendant black fruits, and finally, this is where the bat reference comes in: the clusters of black fruit hanging together look remarkably like clusters of bats roosting

GROWING TIPS

DIFFICULTY RATING: **3**
INDOORS VS. OUTDOORS: Indoors
LIGHT: Moderately bright light to shade. No direct sun
HARDINESS: Not hardy
MOISTURE: Keep consistently moist but not sopping wet. Likes higher than average humidity
GROWING MEDIUM: Well-draining potting mix with 50% extra peat for acidity
NOTES: Slows growth in winter—allow to rest by not overwatering or disturbing roots. Keep above 68°F (20°C) in winter

Did You Know?

The clusters of black fruit hanging together look remarkably like clusters of bats roosting from a cave ceiling or tree branch.

Striking white bracts and whiskers accompany the dark
flowers of white bat plant (*Tacca integrifolia*). Photo by Larry
Mellichamp.

from a cave ceiling or tree branch. Really! (*Tacca chantrieri* is pictured on
page 169.)

While the white bat plant is arguably more striking and beautiful
than the black bat plant, it rarely makes fruit. We have grown the white
bat plant for five years and never gotten a fruit. Therefore, the black bat
plant is more readily available and affordable than its cousin, although
both can be acquired. It is nice to grow both—one for its elegant, jaw-
dropping flowers, the other for its bats.

— 4 —

LOVE PLANTS

Love Is Where the Heart (Shape) Is

Perhaps love is like a window,
Perhaps an open door.
It invites you to come closer,
It wants to show you more.

THE WORDS FROM JOHN DENVER'S SONG "Perhaps Love" remind me of some of the many powerful attributes of love—its inviting nature despite its mystery, the way it opens up a unique universe to each person who is drawn to and engages with it. While nothing we know can compare to the power and complexity of love, I believe that nature does share some of the same characteristics, including an alluring mystery and an invitation to interpret and experience it in our own individual ways. For me and many before me, plants and their unique forms beg to be contemplated, maybe even understood.

Many familiar shapes can be seen in the structures of flowers and plants—stars, half-moons, curlicues, spirals—but no shape draws our interest or warms our insides as much as the heart. When we see a heart it means something to us, perhaps because it is the symbol of love or because it represents what we see as our most vital organ. Today we are merely charmed and intrigued by heart-shaped flowers and plants, but to those in Europe during the Middle Ages the heart shape meant a lot more. It meant life or death, or so they believed. Plants were medicines.

Wax hearts (*Hoya kerrii*) is a bold, vining succulent with suggestive leaves. Photo by Richard H. Gross.

Botany and medicine were one and the same profession, and their study was called herbalism. Herbalists believed that the shapes and structures of plants themselves were clues from God as to the plants' medicinal uses. This idea was known as the Doctrine of Signatures, and herbalists treated ailments with plants or their parts that looked like the affected organ or disease symptom. For instance, if the trouble was with your brain you were given walnuts (think folds of the brain), or if you had kidney stones you were given the kidney-shaped seeds of certain beans. Plants with yellow sap were used to treat jaundice, and—you guessed it—heart-shaped flowers were used to treat ailments of the heart. Nifty idea, but turns out to have zero medical validity. So the use of those heart-shaped flowers may have meant life or death, but there was no telling which one it would be! Lest you think we are so much more enlightened than sixteenth-century Europeans, you can easily find herbal extracts for sale on the Internet whose uses are tied to the shape of the flowers or leaves. Bleeding heart extract seems especially popular and is marketed to those with broken hearts and emotional wounds.

We don't profess that any plants in this chapter have medicinal or spiritual powers, but we are drawn to their names and shapes, and we hope you will be enticed to learn more about them as well. Such is the power of love, in all its many forms.

LOVE-LIES-BLEEDING
Amaranthus caudatus

PLANT TYPE: Herbaceous annual
HEIGHT AND SPREAD: 40 in. × 24 in. (102 cm × 61 cm)

WHILE THE DROOPING, ROPELIKE flower spikes of *Amaranthus caudatus* (Amaranthaceae) are odd attention-grabbers themselves, it is the common name of love-lies-bleeding that brings a sense of macabre melodrama to this old-fashioned annual. We expect a story from this name of mythical or fairy-tale proportions, but no such intact myth is to be found. It seems this name came out of popular imagination. One could imagine the hanging ribbons of red becoming streams of blood flowing from some poor innocent. And the "love" part may have come from mistaking *amar* in the Latin name for "amor." No matter, though. This just means the common name is open for each gardener's dramatic interpretation. We can amuse ourselves when out weeding by spinning our own tales (or tails!) of love and loss around this amaranth.

The scientific name of this plant more realistically describes the actual flower spikes, translating from the Greek *amarantos*, meaning "unfading," and *caudatus*, "having a tail." For the inflorescence is certainly tail-shaped, and when cut from the plant (or left on it for that matter) it remains intact and colorful for months. It is actually this "everlasting" nature that has inspired numerous poets and songwriters to use

GROWING TIPS

DIFFICULTY RATING: **1**
INDOORS VS. OUTDOORS: Outdoors
LIGHT: Full sun
HARDINESS: Not hardy
MOISTURE: Average
GROWING MEDIUM: Average, well-draining garden soil
NOTES: Seeds may self-sow, and a few may come up in subsequent years, but this plant is not generally invasive

Did You Know?
It is actually this "everlasting" nature that has inspired numerous poets and songwriters to use amaranth as a floral symbol of long-lasting, never-fading, even immortal love and devotion.

Love-lies-bleeding (*Amaranthus caudatus*) has striking streaming tails of tiny flowers enhanced by abundant tiny red bracts. Photo by Larry Mellichamp.

amaranth as a floral symbol of long-lasting, never-fading, even immortal love and devotion. Poetry and myth emerge again.

The Incas and Aztecs saw beyond this plant's symbolic nature, straight to its tiny but numerous seeds. *Amaranthus caudatus* (and other related species) were staple crops for these civilizations. Before any of us pass judgment as wheat-, corn-, and rice-centric people, let it be known that amaranth has higher nutritional value, especially in terms of protein and the amino acid lysine, than any of those European, Middle Eastern, or Asian grains. *Amaranthus caudatus* is known as *kiwicha* in the Andes region of South America and is an important food there today. Despite its high nutritive value, it will probably never supersede the cereal grains grown on a massive scale across the globe. Existing technology, machinery, and agricultural systems are already geared toward those crops, and it would take a significant shift to create new technologies to support the harvest and processing of amaranth on a large scale. However, world agriculture does have its eye on expanding the use of amaranth in poor, rural areas where subsistence farmers could easily learn to grow and harvest this tough yet beautiful crop. The plant may develop yet a different story in such places—not so much bleeding as feeding.

LOVE-IN-A-PUFF
Cardiospermum halicacabum

PLANT TYPE: Herbaceous annual vine
OTHER COMMON NAME: Heartseed
HEIGHT AND SPREAD: 72 in. × 12 in. (183 cm × 30 cm)

SOMETIMES BOTH THE COMMON and scientific names perfectly describe their plant, even if the scientific name is a challenge to pronounce! Such is the case with *Cardiospermum halicacabum*. *Cardio* translates as "heart" and *spermum* means "seeds," clearly referring to the round black seeds, each bearing an unmistakable white heart marking. While I think most of us can put voice to the genus name, *halicacabum* is a bit of a tongue-twister and comes out sounding like a made-up word, perhaps part of a spell or the name of a creature from a Harry Potter book. Actually the name is derived from Greek and means "salt barrel," referring to the rounded, inflated seedpod. A complete description of the fruit and its seeds is all in one scientific name.

The common names are equally descriptive. When showing this vining annual plant to children, or adults for that matter, I tell them it is called love-in-a-puff, which usually elicits a head tilt and furrowed brow (body language for, "What are you talking about?"). I pluck a 2 in. (5 cm) diameter, papery

GROWING TIPS

DIFFICULTY RATING: **1**
INDOORS VS. OUTDOORS:
  Outdoors
LIGHT: Full to partial sun
HARDINESS: Not hardy
MOISTURE: Average
GROWING MEDIUM: Average,
  well-draining garden soil
NOTES: Can be grown in a
  container with a trellis
WARNING: Listed as a pest or
  noxious weed in four southern
  American states

Love-in-a-puff (*Cardiospermum halicacabum*) is aptly named: a little, white "heartseed" in an inflated papery pod. Photo by Larry Mellichamp.

brown, balloonlike pod from the vine, hand it to them, and say, "Here is the puff. Can you find the love?" Though they are hesitant at first to break open the inflated fruit, with encouragement they pull it apart and discover three black seeds the size of dried peas. As they examine them and see the perfect white heart, huge smiles cross their faces. Everybody likes a surprise wrapped in brown paper.

I would love to end this story on such an upbeat note but would be remiss if I did not warn potential growers of the vigorously reseeding nature of this species, which is invasive in some areas. Originating from the tropical Americas, it is an annual in zones 5–8, perennial south of that. Be responsible if you want to grow it. Let the kids enjoy picking the "balloons" and finding the hearts, but clean up the whole plant in the fall (let the kids help) and burn or throw away the remains (don't compost). If children are old enough, you can teach them that some plants, even our favorites, can take over other plants. To be a good gardener, we have to keep the plant bullies from ruining the fun for everyone else!

BLEEDING HEART
Dicentra spectabilis

PLANT TYPE: Herbaceous perennial
HEIGHT AND SPREAD: 24 in. × 24 in. (61 cm × 61 cm)

THE COMMON NAME SAYS IT ALL in this case. Who wouldn't want a plant with 2 ft. (61 cm), arching wands bearing perfect pink hearts, hanging from stems as if hung out to dry in the garden? Though simply charming to the seasoned gardener, these hearts are positively squeal-inducing to children and first-time observers. Be sure to point out the flowers on the branch tips that have not yet opened, for they have the most perfect shape, with the large "drop of blood" hanging from the bottom of the heart.

If you are like me, after appreciating the perfect form of the flowers of *Dicentra spectabilis*, the next thing on your mind is tearing one apart. You needn't get too heartbroken if you can't identify what is sepal and what is petal. The components come apart nicely, and it is easy to see how some creative grandmother came up with a story to tell her grandchildren based on the parts of the bleeding heart flower. Naturally it involves a prince, princess, unrequited love, suicide, and regret (what *are* we teaching our children?). As you tear apart the flower, the story is told of the gifts the prince offers the girl to win her love—two magical pink rabbits, a pair of dangling earrings (two sepals, two petals). Of course, she blows him off, so he takes a dagger and stabs himself in the heart (the remaining petals, plus a stamen or the style). The vio-

GROWING TIPS

DIFFICULTY RATING: **1**
INDOORS VS. OUTDOORS:
   Outdoors
LIGHT: Shade
HARDINESS: Zones 3–8
MOISTURE: Keep evenly moist
GROWING MEDIUM: Good,
   well-draining garden soil
NOTES: Goes dormant by early
   summer in all but the coolest
   climates

lent death causes the princess to realize that she was deeply in love with the prince and that her heart will bleed for him the rest of her days. The intact bleeding hearts of the plant bear testament to this. Who can pass up a flower with its own fairy tale?

Look to other species of *Dicentra* for unusual flowers as well. The North American native wild bleeding heart (*D. eximia*) has narrow, heart-shaped flowers, while the native Dutchman's-britches (*D. cucullaria*) bears stalks of flowers that look like white, puffy pants hanging on a clothes-line. Those Dutchmen must have had a reputation, as colonists following the Doctrine of Signatures thought this plant would be good to treat venereal disease!

Bleeding heart (*Dicentra spectabilis*) is a wonderful large garden perennial with fanciful flowers. Photo by Larry Mellichamp.

HEARTS-A-BURSTIN'
Euonymus americanus

PLANT TYPE: Semievergreen shrub
OTHER COMMON NAME: American strawberry bush
HEIGHT AND SPREAD: 60 in. × 36 in. (152 cm × 91 cm)

I THINK WE BOTANISTS ARE LOOKIN' FOR LOVE in some pretty strange places. I mean, the warty red-pink capsules of a shrub cracking open like alien pod-monsters to reveal shiny orange seeds—this is love? Apparently so, as *Euonymus americanus* (Celastraceae) carries the common name of hearts-a-burstin'. Upon closer inspection the heart reference is revealed: the unopened capsules are somewhat shaped like a human heart.

Hearts-a-burstin' is nothing new to North America. In fact, it's a native. As cool as it is with its green-the-year-round square stems, 6–8 ft. (1.8–2.4 m) stature, red fall color, and freaky late-summer seedpods, it is not commonly grown as a garden shrub. Ask around at nurseries that specialize in American native plants and add this conversation piece to your shady garden. As the unopened capsules look as much like little strawberries as hearts, it might be listed under its other common name, American strawberry bush. As the red-pink capsules open to reveal screaming orange seeds, you might question Mother Nature's fashion sense. The birds think it is fabulous, however.

This plant is related to the native bittersweet vine (*Celastrus scandens*) and its robust cousin oriental

GROWING TIPS

DIFFICULTY RATING: **1**
INDOORS VS. OUTDOORS: Outdoors
LIGHT: Full shade to partial sun
HARDINESS: Zones 5–9
MOISTURE: Keep evenly moist
GROWING MEDIUM: Average, well-draining garden soil
NOTES: May form a slowly enlarging clump

Did You Know?
As the red-pink capsules open to reveal screaming orange seeds, you might question Mother Nature's fashion sense. The birds think it is fabulous, however.

bittersweet (*C. orbiculatus*), whose orange capsules open to reveal fleshy orange seeds. Both vines are ornamental during the fall and are widely used as harvest-time decorations, although the latter is an invasive exotic in cooler parts of the South.

Whether you see hearts or just odd-looking seed capsules, *Euonymus americanus* is just waiting to be loved by gardeners with a penchant for the weird.

Hearts-a-burstin' (*Euonymus americanus*) is an American woodland shrub with green twigs and colorful fall fruits. Photo by Paula Gross.

WAX HEARTS AND STRING-OF-HEARTS

Hoya kerrii and Ceropegia woodii

PLANT TYPE: Herbaceous perennial vine
HEIGHT AND SPREAD: *Hoya*, 36 in. × 2 in. (91 cm × 5 cm); *Ceropegia*, 72 in. × 8 in. (183 cm × 20 cm)

WANT TO GIVE YOUR SWEETHEART a living valentine that will remind him or her of your affection year-round? Something that does not destroy your furniture and require housebreaking, that is? Skip the puppy and go for these houseplants. Wax hearts (*Hoya kerrii*) and string-of-hearts (*Ceropegia woodii*), both in the milkweed family (Asclepiadaceae), bear perfect, long-lasting, heart-shaped leaves—reminders of your love and affection.

Hoya kerrii (pictured on page 117) is the larger of the two plants, bearing 2–4 in. (5–10 cm) diameter, tough, waxy green (or variegated) hearts scattered along its thick stems. If happy, this Southeast Asian plant can produce fairly long stems, but these can be trained around a small hoop trellis. The weird-in-themselves clusters of star-shaped flowers are a bonus, looking like they were cut from wax candy. They exude shiny drops of nectar and can smell vaguely of chocolate. Sometimes this plant can be purchased as a single rooted leaf standing straight up in a small pot—a charming (and a bit odd) valentine.

Ceropegia woodii is smaller but packs a lot of love onto its long, trailing stems. If it weren't for the sticky, milky latex that the cut stems exude, one could make

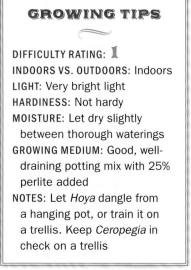

GROWING TIPS

DIFFICULTY RATING: **1**
INDOORS VS. OUTDOORS: Indoors
LIGHT: Very bright light
HARDINESS: Not hardy
MOISTURE: Let dry slightly between thorough waterings
GROWING MEDIUM: Good, well-draining potting mix with 25% perlite added
NOTES: Let *Hoya* dangle from a hanging pot, or train it on a trellis. Keep *Ceropegia* in check on a trellis

Did You Know?

The weird-in-themselves clusters of star-shaped flowers are a bonus, looking like they were cut from wax candy. They exude shiny drops of nectar and can smell vaguely of chocolate.

String-of-hearts (*Ceropegia woodii*) features cute flowers and leaves on a delicate vine. Photo by Richard H. Gross.

a really cool temporary necklace out of this living string-of-hearts. (Don't try this at home.) The cascades of silver-patterned hearts "strung" along the thin stems, hanging from a pot, are fascinating, especially when the strange flowers appear along with the hearts. As with other ceropegias, the flowers are formed like narrow, baroque vessels, colored with dark markings. String-of-hearts flowers are small but intriguing and luckily do not produce a bad odor, as many of this plant's relatives do (for fly-pollination).

Both plants like to dry between waterings and enjoy being pot-bound. Under these conditions they can last for many years. As string-of-hearts ages, it will produce small, marble-sized tubers that can act as propagules to start new plants. So choose your sweetheart carefully when giving wax hearts or string-of-hearts as valentines—the plants may well outlive your affections.

5

ODD INFLORESCENCES

Some Just Don't Make Good Scents

IT'S FRIDAY NIGHT, and the big barn smells of fresh hay. It is decorated with brightly colored flags, and delicious odors come from the refreshment tables. Men and women start arriving, some together, some individually. They mill about, look at each other, wonder what to do. Then the music starts, and the old-time fiddle plays the shoo-fly swing. The people pair up, form a square, and hear the familiar words: "Do-si-do and swing your partner. Now allemande left, and all go home." A time-honored tradition, a quaint but effective way for males and females to meet, enjoy food and drink, talk and make plans, and later, well, who knows what can happen? Where would it all be without . . . a barn?

In the world of flowers and insects, not too much is different in principle from the world of supposedly higher forms of life. Insects have to eat and drink. They need a place to rest. They need a place for males and females to meet, and a place where they can, you know, mate. Where would it all be without . . . an inflorescence? An inflorescence is a large (or small) collection of small flowers arranged in one format or another: spikes, plumes, clusters, cones, or flat-topped stages. These are the nor-

---

Several cyathia of *Euphorbia antisyphilitica*. The colorful parts are glands on the edge of the cuplike bracts, the stamens are yellow, and the three-chambered ovaries hang out. Photo by Larry Mellichamp.

mal types you see every day in a meadow or in a florist shop. Plants arrange their flowers in inflorescences for a variety of purposes: to allow a sequence of blooming, or to fit a large number of blooms into a defined space, or to encourage insects to visit and then leave carrying pollen. Orchids are notorious for having wonderful inflorescences of colorful flowers, but each family has its own general pattern. The most common is the sunflower head—a platform of many flowers grouped together in the middle, with a ring of showy "ray" flowers around the outer edge to help make the whole affair look larger and more inviting. It's the ray flowers that we pick off of daises when we say, "She loves me, she loves me not."

However, not all inflorescences have traditional showy flowers. The odd ones have tiny flowers, often of one sex only, arranged in unusual ways that still provide for cross-pollination, but which are not recognized by us as normal. A spadix is one situation where unisexual flowers are arranged separate from each other on a spike, then placed down inside a sheathlike bract called a spathe. Another unique arrangement is to take extremely small flowers, consisting of only one stamen or one pistil each (no sepals or petals), and gather them into a cuplike nest formed by small bracts. This inflorescence is called a cyathium and is found exclusively in the genus *Euphorbia*. That would not be so much to talk about except that *Euphorbia* is one of the largest plant groups in the world. And here you are, reading a book about plants, so how could we not explain this special flower type? In their own small way (they are typically less than $1/4$ in. [0.5 cm] wide each), cyathia are beautiful as well as functional.

In this chapter we describe some unusual inflorescences that could have gone in the chapter on flamboyant flowers and bracts, except that most of these flowers, as individuals, are not very showy—it is the inflorescence itself that stands out.

VOODOO LILY

Amorphophallus

PLANT TYPE: Herbaceous perennial
OTHER COMMON NAME: Devil's tongue
HEIGHT AND SPREAD: 12–60 in. × 36 in. (30–152 cm × 91 cm)

HERE IS WHERE OUR FASCINATION with the occult intersects with the reality of the plant world, and is often misunderstood. Voodoo is a religious prac-

Did You Know?

The larger-than-life inflorescences are often various shades of speckled maroon but can be white, tan, or cream, mimicking the appearance of decaying flesh.

Voodoo lily (*Amorphophallus bulbifer*) with stinky, cream-colored spathe and spadix. Photo by Greg Allikas.

tice with roots in Africa. It evolved in the Caribbean and Deep South of the United States and deals with efforts to protect oneself and influence the outcome of human events. Followers believe in one God and various spirits that preside over daily life. One can appeal to these spirits through dance, music, singing, chanting, charmed objects, and the use of snakes. The snake spirit is supreme and represents healing knowledge. Thus, to conjure a plant that could be called a voodoo lily you would start with the smooth, narrow leaves of a lily, then add a dark and mysterious (sometimes repulsive) flower incorporating the image of a snake. Presto. You have created a member of the genus *Amorphophallus* (in the arum family, Araceae). These mostly tropical plants are increasingly popular in the United States among those who appreciate the drastically uncommon or weird. *Amorphophallus titanum*, titan arum, is the ultimate member of the genus and is described in chapter 3.

The giant, grotesquely intriguing "flowers" of voodoo lilies are really inflorescences consisting of a spathe and spadix. The spathe is a leafy bract that surrounds the spadix, an aggregation of unisexual, petalless flowers on a spike down inside the spathe. Voodoo lilies are pollinated by carrion flies or beetles, which are normally attracted to rotting carcasses. The larger-than-life inflorescences are often various shades of speckled maroon but can be white, tan, or cream, mimicking the appearance of decaying flesh. The spadix often sticks out well beyond the open spathe, supplying the image of the snake, which makes these plants different from the related jack-in-the-pulpit (*Arisaema* species, see page 138). From this spadix often emanates an odor most foul—again, fly-bait. The flowers usually bloom before the leaves come up, another difference from the jacks.

The flowering period is brief, lasting only a day or two. When these plants are in flower, the smell is powerful, penetrating the area within an acre or so—neighbors will come over to see what has died. The whole flowering structure collapses in four to five days if not cross-pollinated. If pollinated, a conelike aggregation of red-orange berries forms. The plant grows from a massive (bigger than a grapefruit) underground tuber, or corm. It produces a single large leaf that can be several feet

tall and wide (one or more meters) and is divided into many small segments. The leaves and their stalks can be quite beautiful and are often
mottled green and white, or purple. Since the leaf lasts longer than the
flowers, it is the real ornamental part of the plant. The inflorescence is
just for shock value.

Several species are sometimes called corpse flower or arum, and ex-
act identification can be confusing. *Amorphophallus rivieri* (also known as
A. konjac) is by far the most common species grown and is hardy below
10°F (−12°C). It is often available as a pass-along plant—you get one from
somebody who has too many. Another species to seek out is *A. bulbifer*. It
has a creamy white spathe and a beautiful leaf that produces golf-ball-
sized brown tubers directly on the branching leaf that can be removed
and grown into new plants. A related plant is *Sauromatum guttatum*, a
great clump-forming, shade-loving garden specimen
whose compound leaf has a horseshoe-shaped
midaxis. The most beautiful inflorescence of the
group belongs to dragon arum or snake lily (*Dracunculus vulgaris*), a Mediterranean wildflower with a
large, bright purple spathe and several large compound leaves on a 3¹/₂ ft. (1 m) tall stem.

Strangely, if the naked mature tuber of an *Amorphophallus* plant is placed bare on the floor or in an
empty pot in a warm room, it will produce a normal
flower without any roots or soil. Afterward, if you
haven't thrown it away in disgust from the stink, you
can plant it, and it will produce roots from the top of
the tuber and promptly send up a giant leaf. What
could be stranger?

Amorphophallus bulbifer makes a leaf that produces
reproductive tubers. Photo by Larry Mellichamp.

JACK-IN-THE-PULPIT
Arisaema

PLANT TYPE: Herbaceous perennial
OTHER COMMON NAMES: Cobra-lily, Japanese cobra
plant
HEIGHT AND SPREAD: 12–24 in. × 12–24 in. (30–61 cm
× 30–61 cm)

GROWING TIPS

DIFFICULTY RATING: **1**
INDOORS VS. OUTDOORS:
Outdoors
LIGHT: Shade
HARDINESS: Zones 4–9,
depending on species
MOISTURE: Average to moist.
Avoid prolonged wet soils for
most species
GROWING MEDIUM: Good, well-
draining garden soil

WHOA, WHAT'S THAT POKING UP from those
leaves? Why, it looks like cobras getting ready to
strike, or perhaps a troop of strangely helmeted
aliens. There is nothing quite like an arisaema (in the
family Araceae) to surprise even a seasoned gardener
and to bring a sense of mystery and a touch of danger
to a garden bed. Is that due to a combination of the
flower structures' unique forms and coloration, or
the way they actually seem to *lurk* among their foliage? Maybe they just have a magic that we can never
quite explain or comprehend but that makes us want
them even more.

Around 150 *Arisaema* species grow in Asia, North
America, and Africa. The most popular for the garden tend to come from Asia and are often called
cobra-lilies or Japanese cobra plants. The North American jack-in-the-pulpit, A. *triphyllum*, is a favorite wildflower. In all species, the conspicuous part of the
"flower" is really a spathe. The spathe varies in size

Black jack-in-the-pulpit (*Arisaema thunbergii*) comes up
early and produces a long spadix appendage that rests on
the divided leaves. Photo by Paula Gross.

and shape but often forms a hood or cloak and is many times striped in contrasting colors: green, black, white, burgundy, or pink. The true flowers are hidden by the spathe, embedded in the spadix. The spadix is a fleshy axis that in some arisaemas can end in a crazy, long tail that sticks far out of the spathe, either ominously or humorously, depending on your perspective. The fleshy spadixes of arisaemas bear unisexual male or female flowers that are separate from each other, maturing at different times to assure cross-pollination, not self-pollination. If cross-pollination is successful, the gardener is treated to a most interesting cluster of red berries on a stalk come late summer and fall. It is thus necessary to have several plants growing rather close together.

Which ones should you grow? A matter of personal taste, of course, but we have our opinions. Among our favorites that are hardy in average shady garden conditions are white jack-in-the-pulpit (*Arisaema sikokianum*), black jack-in-the-pulpit (*A. thunbergii*), and giant jack-in-the-pulpit (*A. ringens*). The leaves of all are attractive, often with three or more lobes, sometimes mottled, and definitely add their own charm and intrigue to the scene. But this is for you to discover. We wouldn't want to spoil any of the mystery that awaits your affair with arisaemas.

Top left: White jack-in-the-pulpit (*Arisaema sikokianum*) is a very popular and attractive Japanese species. Photo by Larry Mellichamp.

Bottom left: Giant jack-in-the-pulpit (*Arisaema ringens*) has a very large leaf and a constricted opening in the spathe to let flies in. Photo by Larry Mellichamp.

HAND PLANT
Dorstenia yambuyaensis

PLANT TYPE: Herbaceous perennial
OTHER COMMON NAME: Shield flower
HEIGHT AND SPREAD: 12 in. × 12 in. (30 cm × 30 cm)

A NOVICE MAY NOTICE the slightly alien-looking structure among the shiny, average leaves and wonder what it is—maybe a seedpod? But give it to a botanist with a dissecting scope, and you will drive him or her crazy trying to figure out what is going on in there. This inflorescence will be weird even to them.

The hand plant is a member of the fig family (Moraceae). Many plants in this family bear pseudanthia, or

Hand plant (*Dorstenia yambuyaensis*) produces tiny, petalless flowers embedded on an enlarged receptacle edged with lacerate bracts. Photo by Larry Mellichamp.

"false flowers." These are basically a group of really small, inconspicuous flowers arranged tightly such that, together, they look like one big, single flower, as in the familiar sunflower head. Well, dorstenias bear a specific type called a hypanthodium. Zzzzz. Okay, okay! The crazy-looking thing isn't really a flower. It is a flattened top of a stem (receptacle) with fingerlike projections on its margins. Embedded inside that receptacle are tiny, practically invisible, petalless flowers. Dorstenias self-pollinate, but their relatives the edible figs need cross-pollination. Figs have enclosed, saclike hypanthodia, save for a small hole in the end into which tiny wasps crawl and hopefully pollinate the true flowers inside. If they are successful, those flowers turn into tiny crunchy "nuts," and the hypanthodia turn into the sweet flesh of the fig fruit that we eat. The crunch in your Fig Newtons comes from those little nutlike fruits.

EUPHORBIA
Euphorbia

PLANT TYPE: Herbaceous perennial
OTHER COMMON NAME: Spurge
HEIGHT AND SPREAD: 12–60 in. × 12 in. (30–152 cm × 30 cm)

HOW CAN WE POSSIBLY ascribe the label "weird" to an entire genus, *Euphorbia* (Euphorbiaceae), which includes more than 2100 species? Euphorbias are extremely diverse, coming in all shapes and sizes. Some are called spurges, some are called weeds, and some are called desert succulents. But all have a couple of

GROWING TIPS

DIFFICULTY RATING: **2**
INDOORS VS. OUTDOORS: Outdoors
LIGHT: Full to partial sun
HARDINESS: Variable, zones 5–9
MOISTURE: Average
GROWING MEDIUM: Good, well-draining garden soil
NOTES: Some do poorly in hot summers. Many nonhardy, succulent euphorbias are grown as houseplants. The common poinsettia is a nonhardy, indoor, non-succulent euphorbia

Did You Know?
Euphorbias are extremely diverse, coming in all shapes and sizes. But all have a couple of distinct features in common: milky, often caustic sap called latex, and a most unusual inflorescence called a cyathium.

This *Euphorbia* adds colorful bracts to its small cyathia to help attract attention. Photo by Larry Mellichamp.

distinct features in common: milky, often caustic sap called latex, and a most unusual inflorescence called a cyathium. Each cuplike cyathium is made up of tiny bracts and contains tiny flowers consisting of either a single stamen or a pistil and nothing else. Several male flowers, and one female flower, are grouped into each cyathium. The pistil hangs out and matures into a dry capsule, or pod, with three seeds. Cyathia function just like tiny flowers, but they're put together differently. They're really tiny inflorescences.

Some euphorbias have large, showy bracts, like the red Christmas poinsettia (*Euphorbia pulcherrima*) or the white-bracted garden plant called snow-on-the-mountain (*E. marginata*). Desert-dwelling succulent euphorbias do not normally have showy bracts. It is strange that such highly modified flowers work, but they do, and the genus is so large that you have to assume they must work well. Just watch out for the latex.

ARTILLERY PLANT
Pilea semidentata

PLANT TYPE: Herbaceous perennial
HEIGHT AND SPREAD: 8 in. × 8 in. (20 cm × 20 cm)

BOOM! POOF. BOOM! WHOOF. What is going on? A distant battle, with cannons discharging, puffs of white smoke rising in the thick air, and the sound of—wait, there is no sound. Not the 1812 Overture finale at the Independence Day concert or "The Star-Spangled Banner" playing over Fort McHenry. It's a soundless battle, right there on your windowsill. The silent weaponry display is coming from a plant— the artillery plant.

Pilea semidentata (Urticaceae, the nettle family), one of several ornamental species, is a delicate herbaceous ground cover from Panama. It grows all year, forming a mound of green. Periodically throughout the year, especially when the days are about twelve hours long, sets of flat-topped inflorescences with $1/8$ in. (4 mm) flowers appear from near the growing tips and form garrisons of reddish buds. As they mature, they sit waiting, and when the buds are perfectly ready and about to open, they need a spritz of water— like a sudden shower in the jungle—and then poof! bam! boom! The flower buds start to spring open, thrusting puffs of pollen into the air to drift off in a cloud just like the smoke from a cannon. The regiments of flowers burst open with such swift timing that it is hard to follow them across the landscape of the plant. Over here! No, look, back there! Soon the

GROWING TIPS

DIFFICULTY RATING: **1**
INDOORS VS. OUTDOORS: Indoors
LIGHT: Bright light
HARDINESS: Not hardy
MOISTURE: Keep evenly moist. Do not keep too wet
GROWING MEDIUM: Good, well-draining potting mix. Fertilize lightly
NOTES: Cut back annually to rejuvenate growth, or start new cuttings periodically

Did You Know?
The flower buds start to spring open, thrusting puffs of pollen into the air to drift off in a cloud just like the smoke from a cannon.

Artillery plant (*Pilea semidentata*) has clusters of small flowers with stamens that spring open and send up a cloud of pollen resembling cannon smoke. Photo by Larry Mellichamp.

battle is over, and all the flowers have shot their loads. One must then wait for another day when a new set of flowers is loaded and ready to fire. This is the artillery plant's way of spreading its pollen without insects. The pollen is very fine and doesn't go far, but this method of dispersal is the plant's best shot at getting the pollen out there and drifting toward nearby receptive stigmas, perhaps aided by rain droplets that are the triggers to fire.

If you grow this plant and get it into flower and ready to discharge (the buds will be reddish and plump), you need only spray a fine mist and the show will begin in twenty or thirty seconds. Get ready—and don't worry about the smell of gunpowder, holes in the drywall, or waking up the dog.

BEEHIVE GINGER
Zingiber spectabile

Did You Know?
You might even catch real bees visiting your beehive ginger— not because they are nostalgic for those old-fashioned hives, but because the actual flower of the plant is two-lipped, a shape perfectly suited to an investigating bee.

PLANT TYPE: Herbaceous perennial
HEIGHT AND SPREAD: 72 in. × 24 in. (183 cm × 61 cm)

VISIT A MODERN-DAY BEEKEEPER and in her apiary you will see stacks of white, rectangular boxes holding colonies of the social honey-makers. But look in a Winnie-the-Pooh book or at craft store designs featuring bees, and you will see the charming, old-fashioned, cone-shaped bee skep. These are the hives of old Europe, and we still associate them with pastoral pleasure and all that is sweet. So, a glance at the flowering structures of beehive ginger (*Zingiber spectabile*), with their evocative shape and color, brings a smile to most faces.

Of course, the whole thing is an inflorescence—a collection of flowers, and in this case their associated bracts. In fact it is the bracts whose shape, arrangement, and color together form the spike that so charmingly reminds us of old English beehives. The true flowers themselves bloom a few at a time on the conelike inflorescence, and can be imagined as bees visiting the outside of the hive. You might even catch real bees visiting your beehive ginger—not because they are nostalgic for those old-fashioned hives, but because the actual flower of the plant is two-lipped, like a snapdragon flower, a shape perfectly suited to an investigating bee.

The Southeast Asian beehive ginger makes a rather large plant, and it loves good light and mois-

Beehive ginger (*Zingiber spectabile*) has a strongly cone-shaped inflorescence to sequester its small, bee-pollinated flowers. Photo by Ian Odgers.

ture—hence it is not often grown as a houseplant. It can, however, be grown as a warm-weather outdoor plant. To overwinter it, let it go semi-dormant (keep it on the very dry side) and bring it into an unheated sun room for the winter. Bring it out in spring and start watering it well. Even if you don't invest in growing it, you may be able to enjoy it as a cut flower. The stems are increasingly popular in modern arrangements, as the bracts that are the showpiece last much longer than the actual individual flowers. An arrangement of beehive ginger cones may be just the thing to stir the heart of your honey.

6

WEIRD LEAVES

Blades of Glory

N AWFUL LOT ABOUT FLOWERS has been heralded and analyzed in this book. The other plant organs (which do all the daily work, I might add) are so often overlooked and underappreciated. I want to take a stand right now and give leaves their proper due. After all, their primary function is to carry out photosynthesis, and photosynthesis is, like, everything! All life on earth is ultimately fueled by the energy of the sun. Period. Through photosynthesis, autotrophs (organisms that make their own food) absorb energy and miraculously convert it into organic chemical energy in the form of simple sugars. Heterotrophs, like us, who must eat, burn those sugars (and the other organic compounds made from them) to exist, grow, and reproduce. So next time you think leaves aren't worth a second thought, think again!

But this book is not a manifesto of the greatness of plants in general (or is it?). It's about weird plants, and so we should ask, "Where are the weird leaves?" Actually, quite a few types are featured in other chapters: succulent leaves, spines on cacti, the giant leaves of tongue orchids, the outrageous bracts of some flowers, carnivorous leaves, and the leaves of

The leaves of peacock plant (*Calathea*) are beautiful as well as functional. The translucent areas let light through and break up the pattern in order to thwart herbivores. Photo by Maria Pilar Alquézar.

mother fern, which bear baby plantlets along their edges. But what about "normal" photosynthetic leaves? Can some of them be a bit weird as well? Environmental pressure drives adaptation, and leaves are under the constant threat of being eaten by heterotrophs. Herbivores are the next step in the energy chain, and not only can they use the simple sugars in leaves, they can also break down the very structure of plants and extract even more energy thanks to special enzymes and bacterial partners. If the leaves could, they would get up and run away, but they can't. (Although a couple give movement a real try.) Since they can't run, leaves have to fight or hide to protect themselves. They have fought back by evolving a host of strategies to discourage being eaten. One of the most prevalent is manufacturing and stockpiling their tissues with bad tastes, annoying odors, and all-out poisons. They may also wrap themselves in hairs that are unpleasant to the tongues and jaws of herbivores. And in a crowded scene, like a forest or meadow, some may attempt to hide. Often what we see as the beauty in leaves are markings that create camouflage. And strangest of all, some leaves can actually move in an effort to disappear.

This chapter features just a few examples of plants whose leaves have that little something extra. Since we can't just leave well enough alone, we present them here to remind us all that leaves are at least as compelling in their abilities as those old petal-waving flowers.

POLKA-DOT BEGONIA AND PIGGYBACK BEGONIA

Begonia maculata var. wightii and B. hispida var. cucullifera

PLANT TYPE: Herbaceous perennial
HEIGHT AND SPREAD: 12 in. × 12 in. (30 cm × 30 cm)
  or larger

A BEGONIA IS ONE OF THOSE PLANTS we see a lot. In their different forms, they show up as rock-solid annuals for commercial and home landscapes, as florist-shop gift plants, and in bright offices and homes. The genus contains more than 1500 species and can be found throughout the New and Old World tropics and subtropics. So, with so many to choose from, there are bound to be some unique personalities out there. Oh sure, many popular types cover themselves in nice-sized blooms of white, yellow, orange, red, and pink, but what's a little color if you're a flower? Here's an idea—why not put some drama into your leaves? Like the spangled, over-the-top, downright loud costumes of Vegas performers, the leaves of some begonias have forgone the normally understated shades of green. They've pulled out all the stops to make their blooms look like wallflowers and shrinking violets in comparison. These are generally the *Begonia ×rex-cultorum* types, and many cultivars are available from nurseries that specialize in them to keep your taste for clashing colors satisfied. But here we present two begonias that have not been bred to be outrageous—they come by it naturally.

Polka-dot begonia (*Begonia maculata* var. *wightii*) is

GROWING TIPS

DIFFICULTY RATING: **1**
INDOORS VS. OUTDOORS: Indoors
LIGHT: Bright to moderate light
HARDINESS: Not hardy
MOISTURE: Keep evenly moist
  but not soggy during growth
GROWING MEDIUM: Well-draining
  potting mix
NOTES: Can be cut back if they
  grow too large

Did You Know?

The amazing thing about piggyback begonia is that its leaves don't wait for the propagator's knife—they go right ahead and make ranks of little babies along the main veins of their intact leaves.

Polka-dot begonia (*Begonia maculata* var. *wightii*) produces
a fetching pattern of spots. Photo by Larry Mellichamp.

3–4 ft. (0.9–1.2 m) tall and hard to ignore, but not because it has the
flashy color combinations of rex begonias. In the McMillan Greenhouse
at UNC Charlotte, I have rarely seen a person just walk past this plant.
Who knew dots had such power to captivate? There is something about
those perfectly round, sparkly white spots on the long, dark green leaves
that catches the eye. The whole display looks rather unplantlike, more
like the result of what a seven-year-old might do to a leaf given a brush
and white paint. Turn the leaf over, and it looks like the same youngster
coated the back in burgundy. Why would nature so such a thing? We
think spots and patches of white on leaves act to break up the shape of
the leaf on the dark forest floor, mimicking the pattern of sunlight peek-
ing through little gaps in the canopy and thereby helping thwart herbivores from spotting the plant. The sparkliness in the white patches is
also thought to be a light-scattering adaptation to allow the low light that
does reach the leaves to bounce around and have more than one chance
to be absorbed. Furthermore, the red pigment below the green layer can
capture light of higher-energy wavelengths that is found on the dark
jungle floor and funnel its energy to chlorophyll, enhancing photosynthesis. So, it's all about the light, silly.

Piggyback begonia (*Begonia hispida* var. *cucullifera*) does not have colorful leaves or special adaptations for capturing extra light. Its talents lie
in the realm of reproduction. Many begonias can be propagated by chop-
ping their leaves into 1 in. (2.5 cm) pieces and rooting them. The amazing thing about piggyback begonia is that its leaves don't wait for the
propagator's knife—they go right ahead and make ranks of little babies
along the main veins of their intact leaves. It's like the plant has its own
army of clones. The plantlets, or propagules, can detach from the leaf or
wait for the entire leaf to separate from the mother plant. If they land in
a good spot, they will form roots of their own, staking their claim of the
forest floor—and all this without the need for flowers or pollination.
This begonia grows to 2 ft. (61 cm) tall with large, floppy, hairy green

leaves. As the leaves grow and age, the ranks of piggyback riders appear along the veins—like a fuzzy mama possum with her fuzzy little babies, all hitching a ride on her back, until the day comes to shake them off and send them out to make babies of their own.

Piggyback begonia (*Begonia hispida* var. *cucullifera*) produces small would-be plantlets on fuzzy leaves. Photo by Larry Mellichamp.

TELEGRAPH PLANT
Desmodium gyrans

PLANT TYPE: Herbaceous annual
OTHER COMMON NAMES: Dancing plant, semaphore
   plant
HEIGHT AND SPREAD: 36 in. × 24 in. (91 cm × 61 cm)
   or larger

IT IS TRUE THAT PLANTS CAN'T TALK. But they can communicate. Certain oak trees that are attacked by ravaging caterpillars can give off a volatile chemical that is detected by other nearby trees, stimulating them to temporarily produce a greater concentration of harsh leaf chemicals that retard feeding. Colorful petals call out, "Get your nectar here!" to insects and hummingbirds. Ripening cherry fruits turn from green to red to signal to birds that now is the time to feast.

   Desmodium gyrans is a unique example of a plant communicating. Desmodiums are the beggar's-lice of the bean family (Fabaceae), and *gyrans* means "gyrating" or "moving." This plant looks like it is sending signals along the lines of the old flag semaphore system, where a Boy Scout stands and waves two flags in precise motions that send a message as if by Morse code. Each flag position stands for a different letter or number, such as one at six o'clock and the other at nine o'clock to denote the letter *b*. Watch the top of this otherwise unassuming plant, and you will see little leaflets suddenly move in jerky steps—not because you touched them, but all on their own.

GROWING TIPS

DIFFICULTY RATING: **1**
INDOORS VS. OUTDOORS: Indoors
   but may be grown outdoors in
   summer
LIGHT: Bright light but not full
   sun
HARDINESS: Not hardy
MOISTURE: Keep well watered
GROWING MEDIUM: Well-draining
   potting mix
NOTES: Best started new each
   season from seed. Keep
   above 50°F (10°C)

Did You Know?
Watch the top of this otherwise unassuming plant, and you will see little leaflets suddenly move in jerky steps—not because you touched them, but all on their own.

The small leaflets of telegraph plant (*Desmodium gyrans*) spontaneously move up and down in jerking movements as if sending a flag signal. Photo by Larry Mellichamp.

Telegraph plant is a delicate shrub of tropical Asian forests and can become 3 ft. (0.9 m) tall or more. It produces numerous compound leaves, each with three leaflets; the middle one is 3–4 in. (8–10 cm) long, and the other two are only about $1/2$ in. (1 cm) long. At night the large terminal leaflets fold down as if to protect the delicate stems underneath. The remarkable movement occurs in the daytime, when the two smaller leaflets "dance" with quick, jerky motions. There seems to be no pattern as to which leaflets move and when, but if you observe a single leaflet patiently you can see that it moves up—click, click, click—and back down. Some say the jerking motions resemble the tapping of a telegraph key, while others see the movement of the flag signals, hence the common names.

Why does this plant move? One function for the movement might be to attract an insect, if it is trying to lure a pollinator. Conversely, it may be to repel an herbivore, like brushing off a fly. But *Desmodium* moves unpredictably and without direct stimulation, even when no flowers are present. After much observation (though not in its habitat), I have developed a hypothesis. The six to eight small, movable leaflets are always concentrated around the delicate growing tip of the stem, the plant part most vulnerable to insect damage. I believe the leaflets flick sporadically to scare off herbivores, especially those that might lay eggs in the nestlike tip region. In effect the plant is sending a signal that says, "Shoo, bug, don't bother me." I also delight in noticing that at certain configurations, as the leaflets move up and down independently of one another, the array sometimes resembles a fancy Swiss army knife with several blades opened. Secret flag signals, Swiss army knives—maybe I'll rename this the "Boy Scout plant."

SENSITIVE PLANT
Mimosa pudica

PLANT TYPE: Herbaceous annual
OTHER COMMON NAME: Sensitive mimosa
HEIGHT AND SPREAD: 24 in. × 24 in. (61 cm × 61 cm)

PLANTS DON'T MOVE. If the average person thinks they know anything about plants, they know that. Plants don't move on their own—there's no way. Oh, but they do, just not in the way you might be thinking. They don't jump out of their pots and run across a busy street. They don't move fast, but they do have movements; their movements are just normally too slow for us to perceive. Telegraph plant (*Desmodium gyrans*) is just one example. Think also of twining vines, which move their stems in a wide, sweeping spiral pattern as they grow, searching for something to climb up. When they contact something, they narrow the reach of that spiral movement to wrap tightly around their new BFF. This is a specific movement adaptation. Some tropical and temperate plants tend to fold or lower their leaves at night, and if you were to notice you'd think they had wilted. Many members of the bean family (Fabaceae) have compound leaves that fold up at night, presumably to conserve heat and water. Although the leaves move over a period of less than an hour, that's still too slow for us to appreciate.

GROWING TIPS

DIFFICULTY RATING: **1**
INDOORS VS. OUTDOORS: Indoors but may be grown outdoors in summer
LIGHT: Full sun to very bright light
HARDINESS: Not particularly hardy, perhaps zone 9
MOISTURE: Keep constantly moist to wet
GROWING MEDIUM: Average potting mix or garden soil
NOTES: Grows easily from seed. Cut back when too large, and start a new plant from seed every year. Hates winter cold. Keep above 60°F (15°C)

Did You Know?
Touch a single leaflet and it will trigger a wave response such that all the leaflets will fold up, one pair at a time, all down the axis—fwoop, fwoop, fwoop.

Sensitive plant (*Mimosa pudica*) folds up its pairs of leaflets when touched, and produces pink powder puffs of flowers. Photo by Larry Mellichamp.

Lucky for us, a few plants have leaves that move rather rapidly, and during the day.

One such famous plant in the Fabaceae is the sensitive plant (*Mimosa pudica*). (*Pudica* means "shy," by the way.) This sprawling woody shrublet from Central America has prickly stems and bipinnately compound leaves (the leaflets have leaflets). The secondary leaflets, which are about $3/8$ in. (1 cm) long, are arranged in a double row along their axis. When touched, they immediately fold up. Touch a single leaflet and it will trigger a wave response such that all the leaflets will fold up, one pair at a time, all down the axis—fwoop, fwoop, fwoop. (Of course, they're silent, but that's what I *think* they would sound like if they weren't.) They close more rapidly on a warm day than when it's cool. Unlike when you "trip" a Venus flytrap, these leaflets will open back up in ten to fifteen minutes.

Sensitive plant is not carnivorous. So why does it move? Like its name says, it's shy. Sometimes people are shy because they don't want to be bothered or even be seen. Sensitive plant is probably trying to avoid being eaten by a predator. Since the leaves can't run and hide when an insect lands on them, they simply disappear in plain view. If you shake the plant, the whole leaf structure folds down suddenly, making the plant look like a scrawny collection of naked branches. Or maybe the movement is disorienting to an insect's simple nervous system and it jumps or flies away. How does the plant close? Most likely through a sudden loss of turgor pressure in special cells at the base of each leaflet and petiole. Those cells "wilt" completely in a matter of less than a second, causing the leaf to fold.

The genus name and little pink powder-puff flowers of *Mimosa pudica* give away this plant's relation to the attractive mimosa tree (*Albizia julibrissin*). The mimosa tree's leaves don't respond to touch, although they do fold up at night. To amuse yourself with leaves that move, you have to choose the shy cousin over the tall beauty.

AIR PLANT
Tillandsia

PLANT TYPE: Herbaceous perennial
HEIGHT AND SPREAD: 2–12 in. × 2–12 in. (5–30 cm × 5–30 cm)

THE WORLD IS FULL OF AIR. Mostly hot air (and getting hotter). What is air? Mostly it is inert, composed of nitrogen (78%), flammable oxygen (21%), and a teeny little bit of carbon dioxide (about 0.03%, though this is increasing). All organisms need air to breathe because we all need the oxygen for respiration. But plants also need carbon dioxide for photosynthesis. We animals don't get much else from our air. At least we don't get much that is good for us. Maybe that's why we act a little weird from time to time. Maybe there really is "something in the air." Only I don't think it's magic, I think it's pollution.

Back to the plants, though. Bromeliads are a well-known group of attractive-leaved plants found exclusively in the New World. The species in the genus *Tillandsia* in the Bromeliaceae are collectively known as air plants. These tend to be small, tightly tufted, often whitish-looking plants that appear more like giant lichens fallen off tree bark or the skeletons of some delicate sea creatures. They seemingly grow on nothing but air, especially when glued to a twig or a seashell and stuck on your refrigerator. They can survive like this for years. The leaves of air plants dominate, as the stems are barely present and the roots are merely to hold the epiphytic plants to their perches.

GROWING TIPS

DIFFICULTY RATING: **1**
INDOORS VS. OUTDOORS: Indoors
LIGHT: Moderate to bright light
HARDINESS: Not hardy
MOISTURE: Mist every few days. Leaves turn green when scales are moistened. Let dry for several days between waterings
GROWING MEDIUM: No soil necessary. Occasionally spray with weak liquid houseplant fertilizer

Did You Know?
They seemingly grow on nothing but air, especially when glued to a twig or a seashell and stuck on your refrigerator. They can survive like this for years.

Air plants (*Tillandsia* species) with whitish leaves make attractive plants for low-maintenance displays. Photo by Sim Eng Hiang.

Their leaves appear white because they are covered with enlarged scaly hairs. In the humid native habitats of air plants, these scales absorb moisture from the air, and what little nutrients come with rainwater that drips from the canopy above. Because these plants are exposed directly to absorption of water and nutrients from the humid air, pollution hurts them more than most other plants.

The strangest air plant is well known in the southeastern United States as Spanish moss. Neither Spanish nor moss, it is *Tillandsia usneoides*, and it grows well into the tropics. It lacks roots altogether, but the plants grow in great masses that festoon trees in humid regions and swamps. These hauntingly beautiful natural tree decorations make a drive through southern coastal regions unique and unforgettable. People often ask if these air plants hurt the trees. No, they are not parasites, just epiphytes, as are all bromeliads. But they, and other species, often proliferate over time in such abundance as to break old, dead limbs from their wet weight after a rain. Those air plants that fall to the ground are doomed to a stagnant death on the damp forest floor, because they all thrive up high where the sun shines, the humid winds blow, and the carbon dioxide powers their growth. And just like us, it's the other things in the air they could do without.

Spanish moss (*Tillandsia usneoides*) festoons trees in the humid coastal regions of the southeastern United States. Photo by Larry Mellichamp.

7

THE PLANT ZOO

Plants are the Strangest Animals

VEN IF WE LOVE DIVERSITY and revel in differences, some part of us feels most comfortable with and gravitates toward the familiar. Humans are animals, and therefore we have a natural affinity toward other animals—higher animals anyway, the ones with fur, teeth, two ears, two eyes, and one nose. These familiar patterns are burned into our subconscious, and we see them everywhere, in everyday items like a rabbit-ear television antenna, a computer mouse, a dog-eared book page. (Some people see images in potato chips and toasted bread, too, but that's another story.) So no one should be surprised that we see so many animals when we look at plants. Even when we don't see animal forms in them, we might still reference an animal in the common name—cat's-whiskers, rabbit-foot fern, horsetail, dogwood, lamb's-ears, goat's-beard, tiger lily, and on and on. The black bat plant (*Tacca chantrieri*) described in chapter 3 may not be as well known but is entirely evocative of its animal namesake; it's actually the fruits in this case that look like plump black bats, hanging from their roosts in haunting clusters.

As we've alluded to elsewhere, plants and animals are really not so different anyway. They face the same challenges of life: gathering energy, growing, competing, and protecting themselves; undergoing sexual

Black bat plant's (*Tacca chantrieri*) clusters of plump fruits look like roosting bats. Photo by Richard H. Gross.

reproduction and dispersing their offspring while providing them the tools necessary to survive; getting through that tough juvenile stage; and entering maturity. Plants and animals just address these challenges in different ways. The more you learn and think about plants, the more they will draw you into the soap operas of their lives. While we may never go around seeing stamens in lampposts and sepals in sports coats, we can come to consider the plant in everyday life and see it as just a little bit more like ourselves than we ever realized.

Presented here are a few beguiling plants, some part of which remind us of animals. You might think a botanist would grumble against such a zoocentric view of plants, but if seeing animals in plants endears them to us, then I say let's see a bear in every cactus, a giraffe in every tree, and a bird in every flower. Everyone should have a little plant zoo in their own backyards. See how many animals you can grow—you'll have no food to buy, no vet bills to pay, and best of all, no need for pooper-scoopers!

PANDA GINGER AND PIGGIES GINGER

Asarum maximum and *A. arifolium*

PLANT TYPE: Herbaceous perennial
OTHER COMMON NAME: Little brown jugs (*A. arifolium*)
HEIGHT AND SPREAD: 6 in. × 12 in. (15 cm × 30 cm)

THESE LITTLE WOODLAND DWELLERS have a secret. Asarums (also known as wild gingers, in the family Aristolochiaceae) come from the temperate regions of North America and Asia. They make attractive, ground-hugging clumps of usually evergreen, charming, heart-shaped leaves, often with silvery markings. If you crush their petioles or rhizomes you will be treated to a spicy smell reminiscent of culinary ginger, although they are unrelated plants. But that's not their secret. To discover their secret, in spring you have to get on your hands and knees, down at their level, and gently push aside their leaves to have a look underneath. If you are like me, you will squeal with delight as you witness a passel of the funniest little—what are those?—flowers lying against the ground.

Panda ginger (*Asarum maximum*) from China and piggies ginger or little brown jugs (*A. arifolium*) from eastern North America are two species whose flowers are particularly squeal-worthy. Panda ginger flowers are almost 2 in. (5 cm) in diameter and have the most striking black and white coloration and a velvety texture, reminiscent of the famous Asian panda. The shapes of the 1–2 in. (2.5–5 cm) long flowers of

GROWING TIPS

DIFFICULTY RATING: **2**
INDOORS VS. OUTDOORS:
  Outdoors
LIGHT: Dappled shade
HARDINESS: Zone 7–9
MOISTURE: Average. Do not
  keep too wet
GROWING MEDIUM: Good, well-
  draining garden soil
NOTES: Both take a while to
  establish. Once established,
  do not disturb. The fleshy roots
  go deep to anchor the plant

Did You Know?

If you are like me, you will squeal with delight as you witness a passel of the funniest little— what are those?—flowers lying against the ground.

Panda ginger (*Asarum maximum*) has black and white sepals born underneath heart-shaped evergreen leaves. Photo by Rizaniño Reyes.

A. *arifolium* are certainly up for interpretation, but my favorite association is to see a little pen full of piglets. The flowers are the right color and basic shape, and if they are lying at the right angle you can see the pig ears flopped over their heads. Some folks see old-fashioned brown jugs, and others see the mouths of baby birds open and begging for worms.

The flowers of all asarums are unusual, whether or not they remind you of animals. They are very thick and waxy with no petals, just three sepals. They usually form a deep cup, inside of which are the stamens

Piggies ginger (*Asarum arifolium*) produces little brown flowers underneath heart-shaped, mottled evergreen leaves. Photo by Paula Gross.

and pistil. The flowers often have a mild musky odor, and are probably pollinated by small flies (gnatlike), though this has not been definitively proven. Add a wild ginger or two to your shady garden and go on a hunt for pandas and piggies with the favorite child, or childlike friend, in your life. Sometimes the littlest things amuse us the most.

COCKSCOMB
Celosia cristata

PLANT TYPE: Herbaceous annual
HEIGHT AND SPREAD: 12 in. × 12 in. (30 cm × 30 cm)

BRIGHTLY COLORED, VELVETY BRAIN TISSUE. Punk rock rooster who has dyed its Mohawk-like crest Day-Glo colors. Popular garden annual. You choose. One or more features of cockscomb (Amaranthaceae) are bound to pique your interest. The range of flower colors in cultivars of this plant—hot pink, electric yellow, scorching red, Kool-Aid orange—is impressive. But in and of themselves, bright colors don't score overly high on the weirdness meter. It's that vaguely creepy, sinuous form of the flower head that pushes cockscomb into circus plant territory.

"Them flowers just ain't right!" Well, botanically speaking that is kind of true. The flattened, contorted form of growth, sometimes referred to as crested, is not considered "normal," although it does occur naturally and not infrequently in some plants. This squiggly fanning out of tissue is called fasciation (not fascination, although it does provoke that reaction). It forms in plants when the apical meristem (growing point) is disrupted by mutation, particular infections, or damage from certain insects. The meristematic cells respond by dividing in a new direction—out-

The inflorescence of cockscomb (*Celosia cristata*) makes a fascinating fasciation. Photo by Larry Mellichamp.

ward—as well as spiraling upward in the normal direction. If cockscomb's flower heads weren't fasciated it would look more like its cousin plume celosia, with nice, conical, elongated, fuzzy spikes.

We don't see more flowers with fasciation in the gardener's seed box because fasciation is usually something that happens spontaneously here and there in individuals and does not affect the genes of that individual. Any plant, really, can develop a fasciated growth. It happens that the fasciation in cockscomb is a heritable trait, meaning it has established itself in the genetics of the plant and is locked into the blueprints. So when you buy that inexpensive packet of seeds, you can be certain there will be a whole flock of punk rock roosters popping up in your garden, wherever you plant them.

Tarantula cactus (*Cleistocactus winteri*) looks like two tarantulas trying to fit into a pot. Photo by Paula Gross.

TARANTULA CACTUS
Cleistocactus winteri

PLANT TYPE: Herbaceous perennial
OTHER COMMON NAME: Golden rat tail cactus
HEIGHT AND SPREAD: 6 in. × 24 in. (15 cm × 61 cm)

THIS IS A CREEPY CACTUS, and not just because it creeps along the rocky ground in its native Bolivia. It is often called the golden rat tail cactus, but we call it tarantula cactus because we grow it in a tall pot where the bending stems look like two large tarantula spiders trying to climb out of the same hole. Creepy. Even as a young plant in a small pot, it is striking with its initially erect, then arching, hairy "legs." Come to think of it, even if you see the stems as golden rat tails, that's creepy too.

This South American succulent was formerly named *Hildewintera aureispina*, with the specific epithet referring to the yellowish spines, but has been renamed *Cleistocactus winteri*. Considering that most of its relatives are tall and white-haired, this one is distinctly different. It is fun to gently rub your hand one way along the thick mass of spines. But be careful—it can still bite. If you get used to thinking of its stems as tarantula legs, you will be amused when those legs sprout showy, salmon-orange flowers for the first time. Under good conditions, with regular water and fertilizer, it will grow several inches a year. It may surprise you just how far it can grow along a flat surface, sneaking between your other pots. Don't worry, though—it can't get out of the pot and start crawling around your house. It just looks like a tarantula.

GROWING TIPS

DIFFICULTY RATING: **1**
INDOORS VS. OUTDOORS: Indoors but may be grown outdoors in summer
LIGHT: Very bright light
HARDINESS: Not hardy
MOISTURE: Water thoroughly. Let dry between waterings
GROWING MEDIUM: Well-draining cactus mix (50% potting soil, 50% perlite or sand). Fertilize monthly in spring and summer, none in winter

Did You Know?
It is fun to gently rub your hand one way along the thick mass of spines. But careful—it can still bite.

BAT-FACED CUPHEA
Cuphea llavea

PLANT TYPE: Herbaceous annual
HEIGHT AND SPREAD: 12 in. × 12 in. (30 cm × 30 cm)
  or larger

WANT TO SEE A HUMMINGBIRD KISS A BAT?
Well, besides in your overly active imagination, you
could witness this scene (albeit from a distance) in
your own backyard by growing bat-faced cuphea
(*Cuphea llavea*). This 1–2$^1/_2$ ft. (30–76 cm) tall Mexican
plant bears a multitude of tubular, dark purple and
orangey red flowers in summer, perfectly designed
for attracting hummingbirds. So where does the bat
come in? In the flower. The image of a bat face can be
seen quite easily when gazing head-on at a mature
flower. It's kind of a goofy cartoon bat, with googly
eyes, big red ears, and tongue, but it is definitely a bat.
As the hummingbirds visit they tickle the noses of
the bat-faced flower.

GROWING TIPS

DIFFICULTY RATING: **1**
INDOORS VS. OUTDOORS:
  Outdoors
LIGHT: Full sun
HARDINESS: Not hardy
MOISTURE: Keep evenly moist
GROWING MEDIUM: Good, well-
  draining garden soil
NOTES: Cut back if it gets leggy

Use your imagination to see the image in the colorful bat-
faced cuphea (*Cuphea llavea*) flower. Photo by Larry Mellichamp.

TAPEWORM PLANT
Homalocladium platycladum

PLANT TYPE: Herbaceous perennial
OTHER COMMON NAME: Ribbon plant
HEIGHT AND SPREAD: 24 in. × 12 in. (61 cm × 30 cm)

EEW! GREEN TAPEWORMS GROWING out of a pot. This is a plant that once you are told is called tapeworm either elicits a step backward and a crinkle of the nose or a pause and then, "Cool." To each his own. Without the suggestion offered by that common name, some folks find the plant attractive, in a slightly postmodern way. Its flattened, segmented stems grow upward, branch, and arch, creating the look of a sculpture made from stiffened green ribbon.

Homalocladium platycladum comes from the Solomon Islands. It has one of those Latin names that doesn't help the cause of getting people to use more Latin names. If you notice, though, the root *clad* is in both the genus name and specific epithet. This refers to a cladophyll or cladode, which is a flattened green stem that has taken over the function of photosynthesis from the leaves. Tapeworm plant actually does bear some true leaves on its new growth, but these soon drop off, leaving just the stems to do the work of making food. This member of the knotweed family, Polygonaceae, does produce flowers at the nodes of

Tapeworm plant (*Homalocladium platycladum*) looks just like its namesake, with flattened, jointed stems. Photo by Paula Gross.

the stems that are characteristically small and pale. Tapeworm plant can grow up to 3 ft. (0.9 m) tall or more. It responds well to pruning and cutting back for rejuvenation. It can also be propagated from 3 in. (8 cm) cuttings—sort of like what the parasite it is named after does. Eew!

CUCKOO FLOWER

Impatiens niamniamensis

Did You Know?

The flowers look more like fantastical creatures with seahorse tails and bonnet-covered heads, all colored in Life Saver green, yellow, and red.

PLANT TYPE: Herbaceous perennial
OTHER COMMON NAMES: Congo cockatoo, parrot impatiens
HEIGHT AND SPREAD: 12 in. × 12 in. (30 cm × 30 cm)

SEE THE TROPICAL BIRD IN THE FLOWER? Don't feel deficient if you don't. I don't either. The only common name for *Impatiens niamniamensis* that I can relate to is cuckoo flower. Not cuckoo the bird but cuckoo the adjective, as in wacky, nutty, or all mixed up. The flowers look more like fantastical creatures with seahorse tails and bonnet-covered heads, all colored in Life Saver green, yellow, and red. But how do you put that in a common name? Better stick with seeing flocks of brightly colored birds in this eastern African plant.

While the flowers of cuckoo flower are uniquely strange, they do have a feature that is shared by most of the 900 or so species of *Impatiens*—a floral spur. Spurs are nectar-filled pockets formed by sepals or petals that protrude from the backs of some flowers. They are great identification characteristics for botanists, but what does a plant care if some human puts a name to it? Spurs are a way for plants to sequester their nectar so as to only allow appropriate pollinators with long tongues (butterflies, moths, sunbirds, hummingbirds) to access the reward for visiting. The cuckoo flower's spur is curled under, seemingly making that nectar even harder to reach. The plant itself is somewhat succulent and can grow to about 3 ft. (0.9 m).

It is hard not to wonder about the specific epithet *niamniamensis*. Niam-Niam is an old label for the Azande tribe of Sudan, Congo, and the Central African Republic. There is a species of parrot called the Niam-Niam parrot, and the plant does come from those regions in Africa. If you want to really see a parrot in a flower, though, the place to look is in a Southeast Asian impatiens: *Impatiens psittacina* (literally "parrot impatiens"). It is only mentioned here, not illustrated, as the plant is unavailable outside its native range. But do look for an image on the Internet. You will be astonished, as no powers of imagination are needed to see the bird in its flower. Until that botanical wonder can be had, enjoy growing the cuckoo flower and letting your imagination take flight.

Cuckoo flower (*Impatiens niamniamensis*) incorporates the colors of some tropical cuckoo birds and has a strongly curved nectar spur. Photo by Larry Mellichamp.

GOLDFISH PLANT

Nematanthus

PLANT TYPE: Herbaceous perennial
HEIGHT AND SPREAD: 12 in. × 12 in. (30 cm × 30 cm)

WHICH WOULD YOU FEEL WORSE ABOUT forgetting to have someone take care of while you were on vacation, your goldfish or your goldfish plant? Although I have never had a fish tank, I hear that goldfish can go a pretty good while without eating, and I know goldfish plant can take some drying out. I guess it just depends on how deeply your affections lie, and whether you feel worse flushing your pet down the toilet or laying it to rest on the compost pile. Maybe I'd better switch gears. Goldfish plant is like having a little snapshot of an aquarium in a hanging basket. Scattered among the waxy-looking stems and leaves are lots of little goldfish, all fat and orange with their lips puckered up in that goofy, fishy way.

Several species of *Nematanthus* are referred to as goldfish plant. Don't let that throw you—if its flowers look like goldfish, it's a goldfish plant. Goldfish plant belongs to a rather large family of plants, the Gesneriaceae. The gesneriad family contains more than 3000 species, mainly from tropical and subtropical regions of the world. Perhaps the most famous houseplant in this family is the African violet. Several gesneriad would earn the title "weird," including lipstick plant,

The flowers of goldfish plant (*Nematanthus*) are shaped like fish with little mouths. Photo by Paula Gross.

Aeschynanthus radicans, whose unopened flowers look like a tube of red lipstick, uncapped and ready to color the lips of a 1950s movie star. Once open, they are brilliant red tubular flowers, a perfect lure for humming-birds. Goldfish plant flowers, with their narrow "mouths," orange color, and nectar-filled bellies, are likewise hummingbird bait. So forget the goldfish—go with goldfish plant. Hang it outside in the summer and get two pets for the price of one—goldfish and hummingbirds.

UNICORN PLANT
Proboscidea louisianica

Did You Know?

As the green seedpod matures, it turns brownish and tough. It eventually splits open right down the middle of the trunk or horn, producing two diverging, curved spines that can latch onto the leg of a desert antelope or deer.

Unicorn plant (*Proboscidea louisianica*) has beautiful large flowers and curved green seedpods on a coarse plant that is sticky to the touch. Photo by Larry Mellichamp.

PLANT TYPE: Herbaceous annual
OTHER COMMON NAME: Devil's claw
HEIGHT AND SPREAD: 24 in. × 24 in. (61 cm × 61 cm)

THIS PLANT HAS THREE animal references among its names, if you count the devil as an animal. *Proboscidea louisianica* (Martyniaceae) is its Latin name, and not an entirely unique one at that. You see, the genus *Proboscidea* is a genus of animal as well—elephants! How did a plant get the same genus name as the elephant? Obviously they are not related, nor do they come from the same region—the elephant is African, the plant comes from southwestern North America.

It's because of appearances. *Proboscidea* comes from the Latin *proboscis*, meaning "trunk" or "horn."

The animal certainly deserves to be named after its famously long and prominent trunk. But the plant? When the seedpods (fruits) are forming, before becoming fully ripe, they have a long, curving projection from the end of a fattened, elliptical, head-shaped body. They are reminiscent of the trunk coming from an elephant head (minus the big ears). It is this single, but stout, long projection that also earns it the name of unicorn plant. So, what about the devil? Doesn't he usually have two horns? Does he have claws? The devilish character of the pods of this plant emerges upon their maturity (like Regan in the Exorcist). As the green seedpod matures, it turns brownish and tough. It eventually splits open right down the middle of the trunk or horn, producing two diverging, curved spines that can latch onto the leg of a desert antelope or deer. As the animal runs, trying to shake off this annoying hitchhiker, it scatters seeds from the woody seedpod. Some say the pods are so troublesome that the beast sometimes perishes in the panic, and the seeds germinate with a rather large complement of organic fertilizer. I wonder if anyone has seen this plant growing near animal bones.

Although it does have rather large, pale pink flowers, this plant is grown as a curiosity for its pods. People have fun making things out of them, like funny birds and geometric arrangements. It is a bit sprawly, so you might want to grow it in an out-of-the-way location. Just be sure to watch where you step late in the season, especially if you like to walk barefoot in the grass. As interesting as the pods are, it's probably best to avoid testing the theory on yourself that they can take down a large animal. Stick to using them in crafts.

The dry seedpods of the mature unicorn plant utilize strong hooks to attach to a hoofed animal's leg to facilitate seed dispersal. Photo by Jim Gilbert.

FOX FACE

Solanum mammosum

PLANT TYPE: Herbaceous perennial
OTHER COMMON NAMES: Pig's ears, cow's udder, nipple fruit
HEIGHT AND SPREAD: 36 in. × 36 in. (91 cm × 91 cm)

TAKE A LOOK AT THOSE COMMON NAMES: fox face, pig's ears, cow's udder, nipple fruit. They start out G-rated and end up rated R. The fruits of this eggplant relative are like a botanical Rorschach test. What do *you* see in them? All interpretations are fine, unless you see them as "food," as this solanum is poisonous. The sizable, curious fruits are for entertainment purposes only.

Being part of the genus *Solanum*, in the Solanaceae, means having a lot of company. More than 1500 species of nightshades are distributed worldwide. Many are poisonous to one degree or another, often containing the glycoalkaloid solanine in leaves, stems, and (especially unripe) fruits. Solanine and other related compounds cause vomiting, diarrhea, heart arrhythmia, headache, and paralysis in humans, and yet some very popular food plants are solanums—tomato, potato, and eggplant. The ripe fruits or tubers of these plants do not contain solanine and are perfectly safe, though their leaves are not. If you really want to hear

The fruits of fox face (*Solanum mammosum*) may conjure more than one distinct image in the eye of the observer. Photo by Alvaro Senturias.

some fascinating plant stories, look further into the Solanaceae, which includes the famous "witch plants" henbane, belladonna, and mandrake, as well as tobacco and petunias.

The South American *Solanum mammosum* is grown as an annual from seed in temperate zones and can reach more than 3 ft. (0.9 m) tall in one season. Also, be aware that its leaves produce spines. Once the fruits turn yellow or orange and leaves fade or drop, you can strip the leaves and harvest the stems of the plant with the weird fruits attached. This is how the fruits are sold in cut flower markets, especially in Japan and Taiwan. Then let the interpretations begin. If you like the naughtier side of this plant, you can go a step further and include two other R-rated fruits in your garden: *Capsicum annuum* 'Peter Pepper' and *Asclepias physocarpa* (family jewels plant).

BIRD-OF-PARADISE
Strelitzia reginae

Did You Know?
Strelitzia alba *and* S. nicolai *have flowers that look like ghost versions of the classic bird-of-paradise, appearing in shades of gray, black, and white.*

PLANT TYPE: Herbaceous perennial
HEIGHT AND SPREAD: 48 in. × 36 in. (122 cm × 91 cm)

SOMETIMES I LIKE TO USE an Internet image search to test my notions of what I think the popular consciousness is. For instance, I imagined that if I asked someone to visualize an exotic tropical flower, maybe one out of five folks (from the temperate zones) would think of a bird-of-paradise. So I searched for "exotic tropical flower," and bird-of-paradise (*Strelitzia reginae*) did, at least, make it onto the first page. Predictably, hibiscus appeared several times, and surprisingly so did frangipani (the power of fragrance!). Regardless, the large, bold-foliaged South African bird-of-paradise will always be at the top of my list of exotic icons.

The name "bird-of-paradise" refers to no fewer than four types of birds and four types of flowers. Most of the birds with this name have magnificent plumage and often bright colors, especially yellow, orange, and blue. Some have very elaborate mating rituals—why have all that plumage if you aren't going to use it? The bird-of-paradise plant does the same thing with its tangerine-orange sepals and cerulean blue petals, only it attracts sunbirds to carry out the

With its striking flowers and large leaves, bird-of-paradise (*Strelitzia reginae*) presents the classic image of a tropical plant. Photo by Larry Mellichamp.

cross-pollination that will eventually result in baby birds-of-paradise. The flowers emerge sequentially from a horizontally held spathe whose shape reminds one of a bird's head with a long beak. The flowers form the bright crest of "plumage" from the top of the bird's head. Each flower lasts a few days but is generally only ready for pollination for one day. At any time, the spathe holds three to five flowers in various stages of maturity.

While there are actually five species of *Strelitzia* (Strelitziaceae), *S. reginae* is by far the most widely available and grown. It makes an impressive specimen, reaching 4–6 ft. (1.2–1.8 m) tall, and its elegant, upright, paddle-shaped leaves lend a tropical look, even when the famous flowers are not in bloom. *Strelitzia alba* and *S. nicolai* have flowers that look like ghost versions of the classic bird-of-paradise, appearing in shades of gray, black, and white. Curious as they are, those species grow to 15 ft. (4.6 m) or 30 ft. (9 m), respectively—not exactly houseplants. But who am I to say just how far you want to go to achieve the ultimate in exotic, tropical flowers?

— 8 —

PRICKLY PLANTS

The Protection Racket

THESE PLANTS ARE IN THE PROTECTION RACKET, but unlike the gangsters of the Prohibition era, they are not out to offer you a false sense of safety in exchange for your money. They deliver what they promise—you touch them, they touch you back. Most of these "thugs," however, are harmless unless attacked first. They don't belong to gangs. They are loners just trying to protect themselves and assure the success of their species.

These plants are using the latest technology in the plant world—"latest," of course, in evolutionary terms meaning the past few million years. Over that time, they have perfected the use of sharp modified organs for protection against being eaten by an increasingly clever hoard of hungry herbivores. They have to be careful, though. They can't be *too* threatening, because they still need animals, like birds and mammals, to disperse their ripe fruits and seeds. The plants just want to be left alone until their fruits mature. Then the birds can come pick the ripened berries, and the mammals can come get the full-grown nuts and pods off the ground.

Plants can use both their leaves and stems as weapons of mass protection if they are modified appropriately. In everyday talk among gar-

---

The monkey puzzle tree (*Araucaria araucana*) is an ancient conifer whose branches are covered in persistent, sharp, leaflike needles. Photo by Larry Mellichamp.

deners, all the terms for narrow, sharp outgrowths (thorns, spines, prickles, stickers, barbs, teeth) get used somewhat interchangeably. In 1994 I published an article in the *Cactus and Succulent Journal* entitled "Are You Stuck on the Fine Points of Sharp-Object Nomenclature?" wherein I explained the actual botanical difference between thorns, spines, and prickles. Thorns are modified stems, like those seen on hawthorns and young citrus. Spines are modified leaves, as in cacti and barberries. Sometimes just the tips of leaves or their margins can have sharp, pointed outgrowths, such as those of hollies and century plants. Prickles, on the other hand, are neither stem nor leaf but simply sharp outgrowths of the epidermis, as in roses and brambles. They are probably modified hairs. One extreme example is the southeastern United States native needle palm (*Rhapidophyllum hystrix*), which produces dartlike needles more than 12 in. (30 cm) long from its trunk to protect its fruits. These are vicious if you reach into the heart of the plant, because they are so thin as to be nearly invisible to humans. I wonder what animal these palms are protecting themselves from?

Though equipped with sharp weapons, most of these plants will not harm you as you walk through the garden. People and animals have learned to keep their distance, though some of these plants are so fascinating in their outfitting that we can't help but move in for a closer look. Take *Poncirus trifoliata* "Flying Dragon," for instance—get too close and it may appear to reach out and grab you with its ferocious curved thorns. You may get pulled into a situation you hadn't bargained for.

MONKEY PUZZLE TREE

Araucaria araucana

PLANT TYPE: Evergreen tree
HEIGHT AND SPREAD: Up to 30 ft. × 18 ft. (9 m × 5.5 m), usually smaller

WHO CANNOT WANT TO KNOW MORE about a tree with the whimsical common name of "monkey puzzle"? And once you see it, I think you will agree that *Araucaria araucana* is fabulous, unique in form and texture. The common name alludes to the theoretical difficulty a monkey might encounter in figuring out how to climb this tree to get the edible seeds in its large cones. Since monkeys are not associated with the tree in the wild, it is more likely that this ancient "living fossil" was trying to thwart herbivorous dinosaurs from devouring it with its abundance of stiff, sharp leaves.

The whole appearance of the tree does have a *Lost World* flavor—it just looks like it belongs in a Jurassic landscape. The whorled branches are long and sinuous, covered in very closely spaced, sharp-tipped, triangular leaves up to 2 in. (5 cm) long. I suppose these would be more appropriately called fat needles, since the tree is a conifer of an ancient group related to pines. Each needle may persist for more than a decade. The male and female cones are produced on separate trees. The female cones, covered with sharp scales, take eighteen months to mature. They then disintegrate, scattering the edible seeds onto the ground.

GROWING TIPS

DIFFICULTY RATING: **3**
INDOORS VS. OUTDOORS: Outdoors
LIGHT: Full to partial sun
HARDINESS: Zones 6–7
MOISTURE: Avoid keeping too wet in summer
GROWING MEDIUM: Well-draining garden soil. Does not like hot, dry soils
NOTES: Does not do well in summer heat

Did You Know?

The common name alludes to the theoretical difficulty a monkey might encounter in figuring out how to climb this tree to get the edible seeds in its large cones.

The genus name *Araucaria* comes from the Araucano people of Chile, who collect the seeds for food.

The only place to see monkey puzzle in the wild is in the mountains of central Chile and adjacent Argentina, usually above 3500 ft. (1067 m). It is the hardiest member of its genus, and hence it can be grown in temperate regions. It seems to do well in southern England and certain European gardens, southeastern Australia, and New Zealand. The only place you can grow it well (with emphasis on *well*) in North America is the coastal Pacific Northwest and a few areas in the Northeast, like Cape Cod. To thrive it needs cool summers as well as good sun. While it prefers rich volcanic soil (can you buy that in bags?), it will tolerate any well-

The monkey puzzle tree (*Araucaria araucana*) presents flat, sturdy, spine-tipped needles for protection. Photo by Larry Mellichamp.

draining soil. It was once native in North America but disappeared with the dinosaurs at the end of the Cretaceous. If you have the suitable microclimate and the space to grow it, it makes a riveting addition to the landscape. Plus it may confuse monkeys, and who doesn't love a comically perplexed monkey?

HONEY LOCUST
Gleditsia triacanthos

PLANT TYPE: Deciduous tree
HEIGHT AND SPREAD: 30 ft. × 20 ft. (9 m × 6 m)

THE POSTER CHILD FOR EXTINCTION is the dodo bird, which was gone from the face of the earth by 1700. Oh, but what good was it anyway? It was an ugly, flightless bird, apparently inedible, and lived only on the west Indian Ocean island of Mauritius. Human activity wiped it out in just one hundred years. Now we know the dodo did have value as it was likely the sole seed disperser (by swallowing and softening the hard seed coat) for the unique calvaria or tambalacoque tree. Since the dodo disappeared from Mauritius, no new trees have germinated, and the old trees continue to live but are dying out with no seedlings being established. Fortunately, a researcher found a modern surrogate—calvaria seeds eaten by and passed through a large turkey are softened and prepared for germination. So the calvaria tree has been saved. This example shows two things: first, a not-so-rare one-to-one relationship between two organisms, and second, our lack of understanding about what can happen when you change just one small, seemingly insignificant environmental factor.

The honey locust (*Gleditsia triacanthos*, meaning

GROWING TIPS

DIFFICULTY RATING: **1**
INDOORS VS. OUTDOORS:
  Outdoors
LIGHT: Full to partial sun
HARDINESS: Zones 4–8
MOISTURE: Average
GROWING MEDIUM: Good, well-draining garden soil
NOTES: Usually occurs naturally in old fields and disturbed areas

Honey locust (*Gleditsia triacanthos*) has vicious thorns on its trunk but seemingly no enemies or partners. Photo by Paula Gross.

"three-thorned," Fabaceae) is a native of the eastern United States. In the wild it produces a large, twisted bean pod as much as 1 ft. (30 cm) long that contains several small seeds embedded in a sweet, sticky pulp. The bean pods, which fall when ripe, are clearly designed for a large mammal to consume whole off the ground, swallow, digest, and eliminate, with the seeds intact and softened, as the animal roams throughout its range. What is weird is that the tree produces numerous, very large, sharp, branched thorns up to 1 ft. long all along its trunk and branches. What in the world is it trying to protect itself from? It seems to have no modern animal disperser, except perhaps the occasional deer or opossum that might accidentally eat the fruits. Such large thorns are not needed to deter these small animals from climbing into or chewing on the tree. The hypothesis engendered in the late 1970s by ecologist Daniel Janzen is that these fruits were eaten mostly by what are now extinct large mammals that roamed North America during and just after Pleistocene glaciation, which ended some 12,000 years ago. Among these Ice Age mammals were giant sloths, camels, elephants, and giant beavers, and the only remnant of their glory is the American bison. What happened to these animals? While the climate did change, likely they were hunted to extinction by early humans, who came over from Asia via Alaska during the late Ice Age, about 14,000 years ago.

The honey locust today, with its hardly functional fruits and useless giant thorns, is an anachronism, something left over from an earlier time and context. It lives only because humans and modern surrogates keep it going on the fringes of nature. Other examples of modern American tree fruits that belong to an earlier time include Osage orange, Kentucky coffee tree, and giant mossy-cup oak. Realize that all aspects of the life cycles of plants and animals fit in as interlocking parts of a whole, and when a gap occurs, success is hampered. Shouldn't the honey locust be extinct, too? Should we wish we still had dodos? What would we do with them? What would we do without them?

FLYING DRAGON
Poncirus trifoliata "Flying Dragon"

PLANT TYPE: Deciduous shrub
HEIGHT AND SPREAD: 8 ft. × 4 ft. (2.4 m × 1.2 m)

RARELY DO YOU HEAR when walking by a plant, "Don't get too close or it may bite!" The flying dragon may not have teeth, but it has some monstrously long, curved thorns, and once you become entangled in them, they may feel worse than the grip of a determined Doberman. The stems, branches, leaves, and those impressive thorns are all greatly contorted in this variant of the common trifoliate orange (*Poncirus trifoliata*), a member of the citrus family (Rutaceae). The contorted variant originated in Japan and has been around a long time. It is tough in the landscape and very hardy for an orange. In fact, flying dragon is a plant of choice in some areas of Florida for grafting commercial orange varieties onto its very hardy rootstock.

This plant is not strictly a cultivar but a recurring variation of seedlings of *Poncirus trifoliata*. A strict cultivar would be reproducible only by making cuttings of the original, but flying dragon comes true from seed a large percentage of the time, as long as the mother plant is a flying dragon and not just an ordinary trifoliate orange. This means the genes for crookedness are inherited from the mother's cytoplasmic genes.

In all ways except the contorted growth form, flying dragon is like the wild-type trifoliate orange, hav-

GROWING TIPS

DIFFICULTY RATING: 1
INDOORS VS. OUTDOORS: Outdoors
LIGHT: Full sun or very light shade
HARDINESS: Zones 7–9
MOISTURE: Average
GROWING MEDIUM: Good, well-draining garden soil
NOTES: Prune carefully as necessary to keep from growing out of bounds or becoming too thick

Did You Know?
The flying dragon may not have teeth, but it has some monstrously long, curved thorns, and once you become entangled in them, they may feel worse than the grip of a determined Doberman.

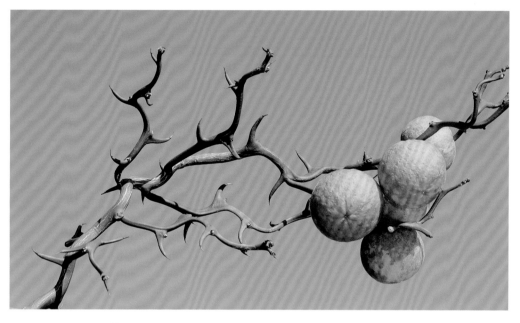

The distinctive curved thorns of flying dragon (*Poncirus trifoliata* "Flying Dragon") stand out in the winter landscape. Photo by Larry Mellichamp.

ing three-parted compound leaves and very fragrant, showy, white spring flowers. It even produces identical fragrant, fuzzy, golf-ball-sized "oranges" in the fall that make nice table decorations, even if they are too bitter to eat. The straight species is rarely grown in the landscape anymore. It forms a dense clump of growth up to 15 ft. (4.6 m) tall and was used in the Old South as an impenetrable hedge, perhaps to keep farm animals in bounds. So, why do I want to grow flying dragon if it is just as thorny as the wild type? Because the gothically elegant contortions of its branches and thorns create one of the boldest silhouettes in the winter garden. The stems of the plant remain green for many years, adding to its wintertime appeal. Plant it where you can see it, perhaps backlit with the winter sun—but not where you might accidentally walk into it, or you might never walk back out!

WINGTHORN ROSE
Rosa omeiensis f. *pteracantha*

PLANT TYPE: Deciduous shrub
HEIGHT AND SPREAD: 36 in. × 36 in. (91 cm × 91 cm)
   or larger

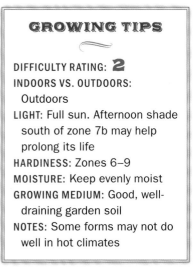

GROWING TIPS

DIFFICULTY RATING: **2**
INDOORS VS. OUTDOORS:
  Outdoors
LIGHT: Full sun. Afternoon shade
  south of zone 7b may help
  prolong its life
HARDINESS: Zones 6–9
MOISTURE: Keep evenly moist
GROWING MEDIUM: Good, well-
  draining garden soil
NOTES: Some forms may not do
  well in hot climates

EVERY ROSE HAS ITS THORNS. But every rose doesn't have thorns like these! Bloodred and so large as to obscure the actual stem, these prickles (sorry, they are not technically thorns) are one of a kind. Nothing could ignore the warning put out by this rose. To the human eye, however, it is stunningly beautiful, especially when backlit.

This remarkable species rose is from well-known Mount Emei and vicinity in the Sichuan Province of China. It may be officially known as *Rosa sericea* subsp. *omeiensis* f. *pteracantha*—this means it is the wide-prickled variant of a subspecies of *R. sericea*—but good golly! It has been more reasonably recognized in *Flora of China* as *R. omeiensis* f. *pteracantha*. Some folks even just refer to it as *R. pteracantha*. Talk about botanical splitting, lumping, and indecision! Sometimes the "science" of taxonomy threatens to make people just throw up their hands and go back to common names. (But as a good botanist, I shouldn't say that.)

As dramatic as wingthorn rose is, it seems to be short-lived, at least for us here in zone 8. And the red color of the prickles does not persist into winter, although the form is still there and makes an interesting winter garden texture. To get lots of new growth in the spring, prune in late winter. It does produce flowers, single white ones (like a wild rose) that have a simple charm. Plant it where it will pick up the morning sun from behind, for the glow of the backlit new stems is so grand. There is nothing like it in the rose world, nor the plant world for that matter.

The wide prickles of wingthorn rose (*Rosa omeiensis* f. *pteracantha*) look spectacular when lit from behind. Photo by Larry Mellichamp.

BUTCHER'S-BROOM
Ruscus aculeatus

PLANT TYPE: Herbaceous perennial
HEIGHT AND SPREAD: 24 in. × 24 in. (61 cm × 61 cm)

LOOKING FOR A PLANT that will protect your hanging sides of meat, clean your animal stalls, and flog your chilblains? Well, if you are living in Britain in the Middle Ages, have I got the plant for you— butcher's-broom. Since you are a modern reader, you'll have to settle for appreciating this plant for its wickedly stiff and spiny habit and alluring cherry- sized red berries.

If you enjoy entertaining your friends with bo- tanical trivia, you must include butcher's-broom in your shady garden. If you plant it in a prominent place, other gardeners touring your beds will inevita- bly ask what it is, for there is something about its stiff angularity and even, deep green coloration that jumps out at the interested observer. This is when you get to say, "Doesn't it have the most amazing stems?" and wait for the visitor to look at you cockeyed. Let the botany lesson begin.

Those stiff, triangularly tapered, sharp-pointed "leaves" that cover the plant are not really leaves but stems. Really! They function like leaves—photosyn- thesizing, bearing protective, spiny tips—but are cladophylls. If your friends don't believe you, show them the plant when it is in flower. Right in the mid- dle of these cladophylls, flowers are born—some- thing that could never happen on true leaves. The tiny

GROWING TIPS

DIFFICULTY RATING: **1**
INDOORS VS. OUTDOORS: Outdoors
LIGHT: Full to partial shade
HARDINESS: Zones 7–9
MOISTURE: Average. Can take dryness
GROWING MEDIUM: Good, well- draining garden soil
NOTES: Cut out old stems as the clump enlarges

Did You Know?
Those stiff, triangularly tapered, sharp-pointed "leaves" that cover the plant are not really leaves but stems.

Butcher's-broom (*Ruscus aculeatus*) produces sharp, ever-green, flattened, leaflike cladophylls (stems) that bear red berries in winter. Photo by Larry Mellichamp.

purple or greenish flowers of butcher's-broom are produced in winter or spring, and if pollinated, form surprisingly large red berries that mature during the summer. The wild-type species is dioecious, meaning it has separate male and female plants, with berries only possible on the females. Luckily for gardeners, self-pollinating forms are available, such as 'Wheeler's Variety', so you can enjoy the berries without having a harem of female plants needing a male to pollinate them. The red berries stay on the plant through the following winter, hoping to get eaten by birds. For this reason, butcher's-brooms make great winter-interest plants in temperate-zone gardens with mild winters. Florists use them in dried flower arrangements, even though they will draw blood if you are stuck by their spine-tipped "leaves." Finally, you can tell your friends that this plant is closely related to asparagus, and its young shoots were even eaten like asparagus in its native Europe in times past. You just never know what stories those humble plants in your garden have to tell.

DEVIL'S THORN
Solanum pyracanthum

PLANT TYPE: Herbaceous annual
OTHER COMMON NAME: Firethorn nightshade
HEIGHT AND SPREAD: 36 in. × 24 in. (91 cm × 61 cm)

DEVIL'S THORN MAKES ME THINK of one of those demons from *Buffy the Vampire Slayer*. I can imagine this lovely 2 ft. (61 cm) tall plant with elegantly lobed and velvety gray-green leaves adorning the garden. Then, suddenly, bam! Bright orange, nasty, large spines start popping through the leaves in all sorts of unimaginable places—the demon plant reveals itself. Seriously, this plant from Madagascar has managed to put prickles not only above and below its midribs but also along the stems and even on the eventual fruits that form (which are poisonous, by the way). Those must be some darn tasty leaves to require such "Caution! Danger!" warnings to keep the herbivores away.

Surprisingly, the prickles on the living plant are flexible and not too nasty if you brush against them. The first time I grew this curiously attractive plant with its screaming yellow-orange prickles, I thought, "Oh, yes, I will definitely grow this again" (from seed here in zone 7b). Winter came, the plant died, and I eventually got around to cleaning out the bed it was in. Ouch! Those long prickles that were once a little flexible and touchable had turned into firm needles able to penetrate gloves. I skipped a few years before letting this plant back into my garden. My advice: Do

Devil's thorn (*Solanum pyracanthum*) produces attractive orange prickles on the leaves that really make you say ouch. Photo by Elayne S. Takemoto.

grow it, but cut it down and dispose of it before it dies for the winter. Enjoy the spectacle without suffering its wrath.

If you like devil's thorn, you might also be interested in one of its relatives, the South American naranjilla, *Solanum quitoense*. It has a similarly intimidating array of leaf spines (this time purple) and another deliciously evil common name: bed of nails.

9

ORCHIDS

An Obsession with Deception

RCHIDS ARE THE MOST INTRIGUING PLANTS in the world. They have fascinated people for thousands of years—at least back to the first century when the Greek herbalist Dioscorides first recorded the use of local wild orchids to determine the sex of human offspring. These weren't the showy, colorful orchids of the tropics but rather diminutive, terrestrial meadow plants. A power and mystery shone through them nonetheless. Each plant bore two underground food-storage tubers: a plump one and a shriveled one. These tubers had a rather evocative shape and appearance, hence the origin of the name *orchis*, meaning "testicle." As the story goes, if you fed your wife the plump tuber, she'd have a male child (the favored outcome at that time). There was no truth in it, of course; but selling these tubers for good prices led to a lucrative livelihood for a resourceful herb collector, since the desired outcome worked about half of the time.

More recently, in the early nineteenth century, European plant collectors were paid huge sums to bring back live orchids from the tropics and to keep the localities secret so that no one else could find them. It took horticulturists back home a while to learn to cultivate these epiphytic orchids in the greenhouses of England and Europe, but it soon

---

The tropical *Dracula* orchid is one of the scariest-looking to us, but what does its pollinator see? Photo by Paula Gross.

became prestigious to grow and show them. And thus began the orchid craze that rose to great heights during the Victorian era, and which continues to almost the same degree today (as skillfully explained in *The Orchid Thief* by Susan Orlean).

References to orchids abound in literature and movies, many of them wildly fantastic. One of the most obsession-generating tales in orchidom was the early report of a black orchid, which many people subsequently spent many years searching for. If a "regular" orchid can hold so much allure and mystery, imagine what powers a *black* orchid must possess! There is no truly black orchid in the wild, just as there is no fountain of youth or city of gold, but it gave explorers something to look for and talk about, and this resulted in the discovery of many new plants. The search for a black orchid continued into the twenty-first century, but not in tropical jungles, rather in the greenhouses of orchid breeders. The myth has finally become a reality, and in this book we report the first cultivated truly black orchid.

The orchid family, Orchidaceae, is perhaps the largest plant family in the world, with some 50,000 species and countless more man-made hybrids and cultivars. Far and away, most of the cultivated types are tropical epiphytes, which in nature would be found growing directly on tree limbs, rocks, or other places devoid of deep soil. They are not parasites but merely grow on tree branches to get up into the canopy closer to the light, a practice shared by numerous other plant groups in the tropics. These plants take well to humid greenhouse and sunroom culture as long as the unique nature of their aerial roots is understood. They can be grown in pots of well-aerated, chopped fir tree bark or mounted on slabs of rot-resistant wood and grown for years with regular watering and feeding. The trick is to simulate their tropical habitat, where nutrients are slowly released from mats of decaying organic matter that accumulate around the branch-grasping roots. It is interesting to think what orchids would have been called by Westerners if the tropical epiphytic species, which are tuberless, had been discovered and popularized *before* the native terrestrial ones.

The flower is the key to the mystery of what makes a plant an orchid.

Orchids have the most highly modified of any type of flower. They are monocots (one of the two major groups of flowering plants, character-ized by having flower parts in threes and sheathing leaves with parallel veins), so we can use the floral structure of a simpler monocot, like an iris, to explain what makes an orchid flower unique. An iris has a symmetrical flower with three each of sepals, petals, male stamens, and female pistils. So, how is an orchid flower different? First, one of the petals is modified into a lip, or labellum, totally unlike the remaining three sepals and two petals (which often look like each other and may be called tepals). This lip is usually dramatically adorned with sculpturing, knobs, horns, or crests. It may be shaped variously into pouches, tubes, wings, or just a flamboy-ant flag. The male and female parts are reduced from three to one, and with special modifications. The single stamen is fused to the single pistil to form a thick structure called the column. On its tip is the pollen mass, formed in a removable apparatus called the pollinium, resembling the twin saddlebags of a pony express rider. All of this fuss is for the boost-ing of cross-pollination, of course. The orchid must attract a bee or other insect to visit it and then transfer a pollinium to the female stigma of an-other flower. When this is accomplished, the result is a pod that contains thousands and thousands of the tiniest seeds in the plant world. It is curi-ous that orchids went (evolutionarily) for quantity over quality (based on the amount of food stored) in their seeds after all the investment in their elaborate flowers. But once again, they end up luring another organism into helping them prosper. Most orchid seeds form a mutualistic rela-tionship with fungi, whereby the fungal threads act like roots, bringing in nutrients for the tiny seedlings until the plants can photosynthesize and provide some photosynthetically produced sugars to the fungus.

Back to the flowers, though. Orchid flowers have some of the most elaborate pollinator-specific adaptations in the world. The size, shape, colorings, and odor of an orchid flower somehow match the physical needs (including sex!) or scent cravings of a particular insect, often a bee. The flower and insect "fit together," much like a lock and key. While some orchids do offer nectar, in many instances the flower deceives the insect by appearing to offer a reward in the form of pollen or nectar, when really there is none. The poor insect is duped into visiting the flower, picking up pollen, and then unsuspectingly visiting another flower of the same species in a vain attempt to meet its needs, only to fail once more. The bee's instinct to gain sustenance is greater than its realization of the futility of the effort, and the orchid benefits by fulfilling its critical goal of getting cross-pollinated. Some of these elaborate flower-insect relationships were first elucidated by Charles Darwin in his 1862 book, *On the Various Contrivances by Which British and Foreign Orchids Are Fertilised by Insects*. Scientists are still discovering amazing relationships between orchids and their pollinators, and these relationships continue to be written about (and filmed) for our voyeuristic pleasure.

Most hobbyists grow orchids on their indoor windowsills or in small greenhouses. Tropical orchids may be simply categorized by their light requirements and temperature preferences. While most tropical orchids can take fairly high daytime temperatures, they do show preferences for certain nighttime (or occasionally seasonal) temperatures. For example, the common moth orchids (*Phalaenopsis* species) that have become so ubiquitous at garden centers and grocery stores need medium light (bright but no direct sun) and warm temperatures at night, 60°F–65°F (16°C–18°C). Cattleyas like higher light and intermediate temperatures, 55°F–60°F (13°C–16°C), while cymbidiums prefer full sun and cool temperatures in the winter, 50°F–55°F (10°C–13°C). Each tropical orchid will fit into one of these temperature regimes. Tropical orchids may be grown in a variety of substrates including bark and peat, or they may be mounted on slabs of cork bark or other suitable materials. Perfecting the watering schedule is another challenge: a few orchids like it wet, but most like to dry out for a few days between waterings. Good air movement is essential, as are higher than normal humidity and regular fertilizing.

Be warned: success at orchid growing can lead to addiction. The next thing you know you are building a greenhouse, installing a 24/7 temperature alarm, and changing your vacation plans because your *Cattleya* is about to come into bloom. Don't underestimate the power of orchids: they are masters of deception in nature and provokers of obsession in humans.

CHRISTMAS STAR ORCHID
Angraecum sesquipedale

PLANT TYPE: Herbaceous perennial
HEIGHT AND SPREAD: 12 in. × 12 in. (30 cm × 30 cm)

IT'S TOTAL DARKNESS in the Madagascan rain forest. The human eye can see nothing. How can anyone or anything maneuver in such darkness? From our limited human perspective, we find it hard to understand that there could be—wow, one just whooshed by. It's a giant hawk moth with a 4 in. (10 cm) wingspread, visiting the large white flowers of a native epiphytic orchid. This is the amazing Christmas star orchid (*Angraecum sesquipedale*), which like many similar species from Africa is fragrant at night, has a long spur, and is known to be pollinated by hawk moths. This particular orchid, however, has an unbelievably long spur, which likewise requires a hawk moth with an unbelievably long tongue. Would you believe—18 inches? That's a foot and a half long. And that is just what *sesquipedale* means.

Charles Darwin studied the Christmas star orchid, and in 1862, trusting in his growing understanding of the intricate workings of orchid flower pollination, he wrote that he believed there should be a moth with a comparably long tongue to fit this particular species. No one believed him. It was just too fantastic a notion at the time. However, in 1903 such a moth was found and named *Xanthopan morganii praedicta* (as in "predicted") in honor of Darwin's suggestion. In 2005, tropical ecologist Phil Devries was able to make

GROWING TIPS

DIFFICULTY RATING: **2**
INDOORS VS. OUTDOORS: Indoors
LIGHT: Bright light
HARDINESS: Not hardy
MOISTURE: Keep moist during growth but a little drier when roots and leaves are not actively growing. Likes 60% relative humidity
GROWING MEDIUM: Epiphytic orchid mix (50% potting mix, 50% medium fir bark)

Did You Know?
This particular orchid has an unbelievably long spur, which likewise requires a hawk moth with an unbelievably long tongue. Would you believe—18 inches?

Christmas star orchid (*Angraecum sesquipedale*) has the longest known spur, made to fit the longest hawk moth tongue. Photo by Larry Mellichamp.

a video of this moth visiting the orchid, its long tongue whipping forward, appearing to us to be a rather awkward tool of the trade. Pretty specialized, but it works—there's less competition at night.

This instance serves as a classic example of coevolution between two species in a very exclusive relationship: the orchid fuels the moth, the moth pollinates the orchid. Many more examples of two organisms driving each other's evolution can be seen in nature, but probably none so unbelievable as the moth with the 18 in. tongue.

SPIDER ORCHID

Brassia

PLANT TYPE: Herbaceous perennial
HEIGHT AND SPREAD: 12 in. × 12 in. (30 cm × 30 cm)

MOST HUMANS ARE NOT EXACTLY drawn to spiders. In fact, most folks are downright scared by the thought of them, especially the large ones. We are drawn to admire orchids, but what about orchids that look like large, long-limbed spiders? It all depends on perception. We may "ooh" and "aah" over the orchid until someone tells us there is a tropical spider that looks just like it. Then we back up a few steps, as if the flower look-alike possesses some of the repulsiveness of the spider itself.

These spidery orchids include several species of *Brassia*. Their petal lips are large but not overly attractive. In some cases they seem to disappear in the context of the flower as a whole, while the five narrow sepals and petals, banded with color, take on an added significance. These are conspicuously splayed out and may be more than 6 in. (15 cm) long. The hypothesis is that these orchids are mimicking large tropical spiders, who sit on their webs across paths and openings in the jungle, hoping to snare an unsuspecting insect. But what advantage would there be to an orchid to look like a spider?

In the tropics, some large wasps hunt spiders, sting and paralyze them, then take them to their nests where they lay their eggs on what will be live food for the newly hatched wasps. We believe the orchids are

The splayed-out flower parts of spider orchids (*Brassia*)
mimic a jungle spider at rest on its web. Photo by Paula Gross.

hoping to be visited by such wasps, who, in an effort to treat the flowers
as spiders, physically wrestle with them, pick up pollen, and carry it to
the next spider-deception encounter. Pretty sneaky on the part of the
orchid. And pretty scary from our point of view if you have ever seen
these large spiders. So, from the orchid's perspective, since the wasp is
not after pollen or nectar, there is no need to advertise as such. Just sit
there, look like a spider, and wait for the wasp's instincts to kick in. Perhaps the wasp will not be fooled too often, but one has to be fooled only
twice for the orchid to gain its advantage. And there are naïve young
wasps born every minute, eager to exercise their instinct-driven reproductive skills.

MEDUSA ORCHID
Bulbophyllum medusae

Did You Know?
They all bloom at once, during breaks in the rainy season, perhaps so that the delicate flowers will not be disturbed by daily showers.

PLANT TYPE: Herbaceous perennial
HEIGHT AND SPREAD: 6 in. × 6 in. (15 cm × 15 cm)

IN GREEK MYTHOLOGY, Medusa had a horrid mass of snakes for hair, and if you gazed upon her you would turn to stone. The medusa orchid has a powerful, but much kinder, impact upon its admirers. What a wonderful surprise each November—like fireworks on the Fourth of July—to gaze upon the newly opened medusa orchid and feel like you have been hit by a stone because of its striking beauty.

This elegant species from southeastern Asia bears flowering clusters that look like waterfalls. The numerous cascading filaments are the sepals, and there are two to four dozen flowers in each cluster. They all bloom at once, during breaks in the rainy season, perhaps so that the delicate flowers will not be disturbed by daily showers. The flowering episodes last only a week each but are worth waiting for every winter. The tiny flowers undoubtedly have tiny pollinating insects. Perhaps the mass of sepals helps to attract by visual appeal, or physically engages an insect and directs it to follow the threads to their source—the waiting flowers. When I look at them I think, "How do they keep that mop of hairlike filaments straight?" I wonder if Medusa had trouble with that too.

Medusa orchid (*Bulbophyllum medusae*) produces cascades of tiny flowers, en masse. Photo by Paula Gross.

The plant consists of many small growths—each a short stem with one leaf, produced from the previous year's short growth, thus creating an ever-expanding thicket. Each new growth can produce a head of flowers, with them all blooming at virtually the same time, so each year the display increases in intensity. As a bonus, a delicately sweet odor emanates from the flowers. It is truly a joy, not a horror, to behold.

TONGUE ORCHID
Bulbophyllum spiesii

PLANT TYPE: Herbaceous perennial
HEIGHT AND SPREAD: 60 in. × 12 in. (152 cm × 30 cm)

LARGER-THAN-LIFE *Bulbophyllum spiesii* is a rare orchid from Papua New Guinea that boasts perhaps the largest leaves in the orchid family—up to 6 ft. (1.8 m) long. They are extremely thick in texture, even for an orchid, and grow from softball-sized pseudobulbs. A pseudobulb is swollen stem tissue at the base of leaves used for storing extra water, and is a common anatomical feature of many orchids. Pseudobulbs often remain as storage reserves for the plant after the worn-out leaves fall off—in this case, after many years.

In addition to its amazingly large, tongue-shaped leaves, this orchid has rather odd, clustered flowers. Produced in summer, they are dark burgundy-red with leathery sepals cupped around and hiding the smaller lip and other flower parts. While they are generally not considered beautiful, they will attract your attention with their unbelievable rotten smell. Why so stinky? To attract pollinators like bottle flies and carrion beetles. The insects come to the flowers because they look and smell like rotten meat. They lay their eggs there but then find nothing for their maggots to feed on. Another clever deception.

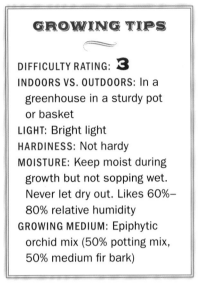

GROWING TIPS

DIFFICULTY RATING: **3**
INDOORS VS. OUTDOORS: In a greenhouse in a sturdy pot or basket
LIGHT: Bright light
HARDINESS: Not hardy
MOISTURE: Keep moist during growth but not sopping wet. Never let dry out. Likes 60%– 80% relative humidity
GROWING MEDIUM: Epiphytic orchid mix (50% potting mix, 50% medium fir bark)

Tongue orchid (*Bulbophyllum spiesii*) produces very large leaves and stinky flowers. Photo by Paula Gross.

Because the plants are rare, they do not serve as effective birth control for the flies in the area, which is a good thing. The lowly maggot has its place in the circle of life—helping to keep the environment clean of carrion.

The leaves and flowers of tongue orchids may seem weird to us, but like everything in nature, these plants are perfectly adapted to their native habitat. They grow as epiphytes, hanging from the crotches of sturdy tree trunks, or sometimes on rocks where organic matter has accumulated. *Bulbophyllum phalaenopsis*, also available in cultivation, is similar but a bit smaller. Why does *B. spiesii* grow so big? Because it can! (Not unlike the egos of humans.) Perhaps it is not just the giant, tonguelike leaves that earn this orchid the cheeky nickname of "Mick Jagger orchid."

TRIGGER ORCHID

Catasetum saccatum

PLANT TYPE: Herbaceous perennial
HEIGHT AND SPREAD: 12 in. × 12 in. (30 cm × 30 cm)

CATASETUMS AND THEIR CLOSE RELATIVES in the subtribe Catasetinae are a group of orchids native to the New World tropics whose flowers have some special features that set them apart. For instance, they produce separate male and female flowers—a most uncommon condition in orchids—and the male flowers are actually more showy and attractive than the females. The several different genera in the group produce significantly different flowers.

A male bloom of *Catasetum saccatum* is 1 in. (2.5 cm) wide, colorful, and fragrant. Its odd, waxy form makes you want to touch it. You reach out and stick your finger on its "mouth." Whoa, jump back—what just happened? You may not have noticed, but you touched a little curved trigger hair, and in a split second the flower popped out its pollinium and attached it to your finger with a pad of hold-fast glue. Go ahead, try to shake it off. It would stick all day if you didn't pry it off (it won't hurt you, though, it's just plant tissue). This is what happens to a male euglossine bee when he enters the flower. He is scared off by the jolt, leaves with this thing on his head, and (hopefully) visits a female flower wherein the pollen mass is transferred to the stigma and effects cross-pollination. An elaborate scheme, but it works.

So, why are the male bees attracted to the orchid

GROWING TIPS

DIFFICULTY RATING: **3**
INDOORS VS. OUTDOORS: Indoors
LIGHT: Very bright light
HARDINESS: Not hardy
MOISTURE: Water heavily in summer. Likes 60% relative humidity
GROWING MEDIUM: Epiphytic orchid mix (50% potting mix, 50% medium fir bark). Fertilize heavily in summer
NOTES: Stop watering and fertilizing in December, as the plant goes dormant in winter and sits dry and leafless

Did You Know?

You may not have noticed, but you touched a little curved trigger hair, and in a split second the flower popped out its pollinium and attached it to your finger with a pad of hold-fast glue.

A male bloom of a trigger orchid (*Catasetum saccatum*) cultivar with sprung pollinium attached to finger. Photo by Paula Gross.

in the first place? These bees lead an unusual life. They make a nest, then hope to entice a female to it in order to mate by bringing in an alluring smell she might like. But they don't produce this pheromone themselves—they collect it from certain orchids! The orchids produce a waxy, perfumed coating, and the bees come to scrape this off the flowers and carry it back to their bachelor pads. Different orchid species produce certain chemical odors that attract certain bee species—again, a closely specific relationship. The odors of trigger orchid have been described by one orchid grower as "a heady combination of spearmint, mothball, floor wax, grapefruit, and maybe cheddar." Apparently there is no accounting for individual taste, whether in human society or nature.

BUCKET ORCHID
Coryanthes macrantha

PLANT TYPE: Herbaceous perennial
HEIGHT AND SPREAD: 12 in. × 12 in. (30 cm × 30 cm)

THIS IS THE RUBE GOLDBERG DEVICE of the orchid world. There is no more complicated orchid flower than that of *Coryanthes*, and perhaps none more beautiful and interesting (see page 16 for a photo of an unopened bud that looks like the Man in the Moon). The several species range from Guatemala to Bolivia in warm lowland rain forests that may receive more than 150 in. (3.8 m) of rain per year. They typically thrive in masses of epiphytic vegetation called ant nests. The ants create an acidic environment, produce organic waste for fertilizer, and protect the plants from herbivores.

Of course it's the unbelievably complicated flowers and how they work that amaze us. The "wings" of the flower are the two side sepals. The "bucket" is the lip, and it is highly modified with a large knob at one corner of the holding arm. This is where a powerful scent is produced that attracts swarms of male euglossine bees (remember them from catasetums) who come to collect fragrance to take back to their nests to attract female bees. The odor may have hints of vanilla, eucalyptus, and other volatile fragrances. (If you are into chemistry, they are benzaldehydes and terpenoids—so there.) An eager male bee comes to visit, begins scraping off the odoriferous coating, slips on the smooth arm, and falls into the bucket

Bucket orchid (*Coryanthes macrantha*) gives a bee a thrill with a spill. Photo by Paula Gross.

where a viscous liquid has been collecting, dripping from the "faucet glands" just above. The drenched bee cannot crawl or fly out, for the walls are slippery and he is wet with sticky fluid. The only way he can escape is to squeeze out the end of the bucket through a small hole made by folds in the lip surrounding the column. In doing so, he passes beneath the pollinium, and the pollen load becomes stuck onto his back. He then dries himself off and flies away in disgust. His memory is short, though, because he quickly visits another flower with exactly the same odor, and does the same thing all over again. Will he never learn? Not really—the lure of sex is too strong. The second time he falls in and squeezes out, he deposits the pollinia on the stigma, effecting cross-pollination.

All of this must be accomplished in the first two days the flower is open, for then the flower is spent. If pollinated, the ovary swells with hundreds of thousands of seeds (the most of any orchid), and the whole ruse has been worth the investment. If pollination does not occur, the plant tries again with a new flower to follow. After all, there's a sucker born every minute.

BLACK ORCHID
Fredclarkeara After Dark 'Black Diamond'

PLANT TYPE: Herbaceous perennial
HEIGHT AND SPREAD: 12 in. × 12 in. (30 cm × 30 cm)

IF YOU SEARCH THROUGH LITERATURE and advertising, you will find many references to "black orchid." It can refer to perfumes, drinks, a movie, novels, resorts, jewelry, a flower, and so forth. It is an alluring concept—one of mystery, longing, passion, and secret desires. This mystique of orchids stems from the myths of long ago created by explorers who came back from exotic tropical locales with tales of their exploits and what they thought they saw but couldn't quite get (like the fisherman's "one that got away"). Perhaps this was to stir the fires of desire so that their benefactors would send them back for more, in hopes of being the first to find the rarest and most elusive specimens.

But now here it is. It's black. It's really black! Did someone finally discover the long-lost mythological black orchid in some hidden tropical paradise? No. I am afraid that the wild, mystical jungle that protects the black orchid does not really exist. Nor does the orchid itself—in the wild, that is. This black orchid was actually created in the horticultural laboratory of the greenhouse. You might think we are breaking our own rule by including a plant that does not exist in the wild. Well, yes and no. This black orchid is the result of a chance combination of genes brought about by a series of hybrid crosses among selections with

GROWING TIPS

DIFFICULTY RATING: **3**
INDOORS VS. OUTDOORS: Indoors
LIGHT: Bright light
HARDINESS: Not hardy
MOISTURE: Water heavily in summer. Likes 60% relative humidity
GROWING MEDIUM: Epiphytic orchid mix (50% potting mix, 50% medium fir bark)
NOTES: Should flower at Christmas. Stop watering and fertilizing in December, as the plant goes dormant in winter and sits dry and leafless

Did You Know?
When prominent orchid show judges first saw Fredclarkeara *After Dark 'Black Diamond', they tried to say that it was just really very dark purple, like so many before it. But in the final analysis, all had to agree that these flowers were truly black.*

Fredclarkeara After Dark 'Black Diamond', the world's first truly black orchid—a man-made miracle. Photo by Fred Clarke, courtesy of Sunset Valley Orchids.

ever darker flowers. It is not a one-of-a-kind freak but a genetic modification that would come true a certain percentage (just not 100%) of the time from seed.

Fred Clarke, an orchid grower at Sunset Valley Orchids in California, spent many years hybridizing dark-colored orchids in the *Catasetum* "tribe" (a group of closely related genera). He used several different genera in the process, including *Catasetum*, *Clowesia*, and *Mormodes*. After raising thousands of seedlings, he finally got one that produced really black flowers—a first in the orchid world. This is a matter of luck, skill, patience, and many trials. The spectacular result of his efforts was named after him: *Fredclarkeara*, a new genus. (The rules of orchid nomenclature dictate that when you use four or more different genera in an orchid hybrid, you have to make up a totally new genus name.) The cultivar name 'Black Diamond' is a horticultural designation of a clonal selection of the best from among many slightly different seedlings, somewhat like the registered name of a pedigree dog. When prominent orchid show judges first saw *Fredclarkeara* After Dark 'Black Diamond', they tried to say that it was just really very dark purple, like so many before it. But in the final analysis, all had to agree that these flowers were truly black. Fred Clarke has propagated his "discovery," and specimens can be had if you would like to test the mystery and darkness of this long-awaited icon.

GONGORA ORCHID

Gongora

PLANT TYPE: Herbaceous perennial
HEIGHT AND SPREAD: 12 in. × 12 in. (30 cm × 30 cm)

THE GONGORA ORCHID FLOWER appears upside down compared to most orchid flowers. This may not even be worth pointing out, considering that we still wouldn't be able to explain why it looks the way it does even it if it were right side up. Except that flowers are produced to best effect pollination from their particular pollinator. Here is yet another case of a strong

A strange *Gongora* flower with sepals flared back looks like a pterodactyl in flight. Photo by Paula Gross.

odor and weird flower attracting the euglossine bee. Gongora orchids grow in dense clumps in the steamy, wet jungles of Central and South America, and were apparently one of the first tropical orchids to be described by Western civilization. The inflorescences hang down and often bear dozens of flowers that last a few days each. They may be dark-colored, patterned, or nearly white, and they emanate intriguing smells similar to those of *Coryanthes* and *Catasetum*.

If we want to be botanically accurate (and why not?), the flowers of gongoras are actually right side up as you look at them. Most other orchid flowers are held upside down when ready for pollination, with their lips at the six o'clock position. This is because the flower stalks twist during development, and this condition is called resupinate. No matter how you look at it, gongora flowers resemble *to us* pterodactyls or some mythical fairylike creature. Their winglike sepals are held back out of the way of the conspicuous lip, while the column with stigma and pollinium curve gracefully downward. The euglossine bee that visits hangs upside down and tries to get the scent off the lip in a process called brushing, making use of specialized stiff hairs on his front legs. This is an awkward position at best, akin to painting the ceiling of a room while balanced on the underside of a light fixture. And this is not an accidental arrangement on the part of the flower. When finished, the bee drops down, falling onto the column below, where he picks up the pollinium on his back. You know the rest of the story by now. Yep, he goes off and does it again. He is not hurt, and it is fun to watch.

DOVE ORCHID

Peristeria elata

Did You Know?
The perfect image of a folded dove is formed inside the cup-like flower, much like a religious icon inside a church niche.

PLANT TYPE: Herbaceous perennial
OTHER COMMON NAME: Holy Ghost orchid
HEIGHT AND SPREAD: 36 in. × 12 in. (91 cm × 30 cm)

WOW! THE DOVE ORCHID is also sometimes called the Holy Ghost orchid—quite a culture-rich common name to live up to. Does this mean *Peristeria elata* is the salvation of the orchid world? No, but this elegant national flower of Panama does evoke much reli-

Dove orchid (*Peristeria elata*) has a lip that perfectly resembles a hand-cupped dove. Photo by Richard H. Gross.

gious symbolism. The specific epithet means "lofty," as in elation, and the genus name is Greek for "dove," the bird that often represents the Holy Spirit in religious paintings. It is not at all difficult to see how the genus got its name. The perfect image of a folded dove is formed inside the cuplike flower, much like a religious icon inside a church niche. The cup is made up of the sepals and two petals. The third petal, the lip, forms the tail and wings of the dove. The column forms the head and beak. Amazing!

What does a bee make of this flower? From our perspective the visiting bee seeking nectar lands on the tail of the dove, which is technically the flower's lip. This lip is not fixed but wobbles on a fulcrum, moving the bee inward. Then the bee flips backward, catches its head underneath the "beak" of the dove, and pops off the cap (dove's head) to reveal the pollinium, which becomes attached to the head of the bee. When the bee visits another flower, the same movement is repeated, and the pollinium, now sticking out like a unicorn horn on the front of the bee's head, is thrust up into the stigma. This whole affair is a bit tricky, though, since the stigma consists of a narrow slit, and in order for the two pollen masses to fit inside, they must reorient themselves from vertical to flat. I have cross-pollinated the dove orchid flower by hand, and it is very difficult to insert the thin pollen sacs into the tiny slit. There are far more attempts than successes in the wild. Maybe it is just as hard to be a good orchid pollinator as it is to be a good religious figure. But they both keep on trying.

Dove orchid is a robust plant, with pseudobulbs as big as large grapefruits. These store water for the dormant dry season and then send up large leaves more than 3 ft. (0.9 m) tall. The 3–6 ft. (0.9–1.8 m) flowering stalk produces dozens of 3 in. (8 cm) pure white flowers over a period of several weeks. They smell wonderful on close inspection.

BUTTERFLY ORCHID
Psychopsis papilio

PLANT TYPE: Herbaceous perennial
HEIGHT AND SPREAD: 12 in. × 12 in. (30 cm × 30 cm)

MEMBERS OF THE GENUS *PSYCHOPSIS*, including
the butterfly orchid, are often called dancing ladies
because the lowest petal, or lip, is larger and more
flamboyant than the other petals, so the flower looks

Graceful butterfly orchid (*Psychopsis papilio*) appears to flutter in midair.
Photo by Paula Gross.

like a ballroom dancer in a long, ruffled dress moving across the floor, often with arms raised high (the other two petals and one sepal). These are colorful orchids, usually numerous on the spray stem, and often quite fragrant.

The one we call butterfly orchid, *Psychopsis papilio* (syn. *Oncidium papilio*), bears vibrant flowers one to three at a time atop long, willowy, 2–3 ft. (61–91 cm) stems that sway in the slightest breeze. Each flower can be 5 in. (13 cm) tall. This floating movement combined with the narrowed upper sepals and petals that resemble antennae, and the lower three that are flared out and winglike, suggest a large, colorful butterfly. But why should these flowers look like butterflies? Are they pollinated by butterflies and thus hoping to attract would-be mates who will be fooled into carrying the pollen? Is it another kind of deception, perhaps an attempt to misguide a would-be predator? Or is it just a distinctive pattern that is easy for a pollinator to remember? We don't really know.

We do know that this particular orchid seems to beg for interpretation. If you don't tell folks its name, but rather ask them what it looks like, they are inevitably drawn in, study it, and come up with a diverse range of answers, from twirling ballerina to crab to jungle mask. The beauty of orchids is that we can imagine many things as we fit our experiences onto their unconventional flower forms. Yes, we know they are all designed to attract pollinators, but sometimes it is more fun to forget that explanation and just let our imaginations run wild. This must be one reason we admire orchids so much—they reflect or own experiences and allow us to wonder.

10

SUCCULENTS

Saving Water from a Rainy Day

T HEY SAY DOG OWNERS often resemble their pets (or vice versa). What if our favorite plants resembled us? We'd have fat plants, white-bearded and mustached plants, tall and slender plants, gnarly and creepy plants, slow growers and those that grow on you, shy bloomers and those bursting with color and drama, plants that hibernate in winter and those that get lethargic in hot weather, plants that smell bad, and so on. These examples, and many more, perfectly describe the category of plants we call succulents.

They come from the great deserts of the world and are adapted to long periods of little or no rainfall, usually combined with hot conditions. The word *succulent*, in its common usage, means "juicy." Botanically this means that these plants store water in modified and enlarged organs—roots, stems, or leaves. This stored water allows succulents to carry on somewhat normally during the worst of the dry times without going totally dormant. Thus we have root succulents, including underground bulbs and tubers, which die down to the protected, nonphotosythesizing underground storage organ during the dry season. Leaf succulents, such as hens-and-chicks and sedums, have reduced or under-

A variety of succulents of all sizes and shapes on outdoor display at the Huntington Botanical Gardens in San Marino, California. Photo by Larry Mellichamp.

ground stems, so that the vast majority of the plant body is made up of enlarged, water-storing leaves. Stem succulents store water in enlarged (fat!) stems and usually have spines or prickles instead of leaves, as in cacti and desert euphorbias; the green, succulent stems carry on photosynthesis, taking over the function of the absent leaves.

By the way, there is no "liquid water" inside a fat desert plant. Succulents store water in spongy tissues made up of enlarged cells. I'm afraid it's not like in cartoons where the thirsty cowboy finds a barrel cactus, taps it with a spigot, and out flows water into his waiting glass. At best, you'd be able to cut one open and squeeze moist, slimy sap into your mouth. It may save your life, but nature's watercooler it is not. Furthermore, some succulents, especially in the euphorbia and milkweed families, have stems and leaves filled with toxic milky sap that is irritating or even dangerously caustic to skin and eyes. One of the worst is the commonly grown houseplant called pencil tree "cactus" (really an African euphorbia), whose perfectly round, leafless stems are about the diameter of a pencil. It is reported that in Africa its sap has been used to brand cattle. (Don't try this at home.)

If you live in a desert and hold water, you will be viewed as a source of moisture for thirsty animals. Consequently, every succulent has a strategy to avoid being consumed for its juices, usually falling into one or more of these three conditions: physical protection in the form of spines or thorns, chemical protection in the form of poisonous sap, or concealment, as in hiding under the sand or among pebbles to avoid detection. Animals aren't the only threat to a desert plant's water stores and health. Intense sun, heat, low humidity, and drying winds are stressful, and desert plants have strategies to deal with them as well. Some grow under bushes or behind rocks, seeking the shade, so to speak. Others have white hairs or white waxy epidermal coverings that reflect sunlight and keep the plant cooler. Then there is the almost universal condition among succulent plants called Crassulacean acid metabolism, or CAM. It was first discovered in the succulent genus *Crassula* when a researcher noticed that the fat leaves tasted sour in the morning and bitter in the evening (way to get intimately involved with your research subject!). He found that the plants make malic acid at night from the carbon dioxide they take in (sour in the morning), then reverse the process during the day to supply intracellular carbon dioxide for photosynthesis (becoming less acidic, or bitter, in the afternoon). This allows them to keep their water-losing stomata (leaf pores) closed during the hot daytime. They open only at night, when it is cooler. What a great system!

The regions that have produced the most diversity of succulents are the deserts of southwestern North America (Mexico including Baja California and adjacent United States), deserts of western South America, and deserts of southwestern South Africa. On the contrary, the cold northern deserts (Gobi and Great Basin), the very dry Sahara, and the vast interior of Australia have produced very few succulents of interest to hobbyists, probably because those areas are just too dry for much of the time. Within the regions that are good for succulents are areas where rain falls only in winter, including Baja California and the Cape region of southwestern Africa; plants from these places grow during the short days of winter when the rains come, and do not adapt well to summer watering. Others come from regions with summer rain and want their water when days are long and temperatures hot; they will rot if given anything more than a monthly sprinkle in the winter. A few succulents are less fussy and will adapt to any scheme that gives them strong sunlight and periodic water. It is important to know where your plants come from if you hope to grow them properly. An overwatered succulent is like a pig at an endless food trough—it doesn't know when to stop and will absorb water until it distends abnormally and comes close to bursting.

Thousands of plants around the world have developed some degree of succulence that allows them to

Crassula rhodogyna has overlapping sets of leaves that make a neat pagoda. Photo by Paula Gross.

survive in deserts. They are in many different, unrelated families. For example, true cacti (Cactaceae) come from New World deserts only. South Africa is home to many desert plants that are totally unrelated to cacti (like some euphorbias, milkweeds, and senecios) but look so similar to cacti in form as to be indistinguishable to the nonspecialist. If it weren't for the differences in their flowers, even botanists would have a heck of time sorting them into their true families. Maybe this is why the average person tends to refer erroneously to all desert succulents as "cacti." Another case of look-alikes occurs with New World leaf succulents like sedums and echeverias, which may look like the unrelated Old World crassulas and haworthias when not in bloom. This phenomenon is called convergent evolution, where unrelated plants come to look alike due to evolving similar adaptations to similar environmental conditions. Nature has been shaped by millions of years of nurture, as it were.

To successfully grow these highly adapted plants outside their native habitats, you have to understand and respect their unique needs. Extreme succulents, such as those with spherical bodies and white hairs or coatings, love heat and bright light. Growing such plants on a shady windowsill, thinking you are protecting them from heat, will not make them happy. They are adapted to heat, light, good air movement, and arid conditions, and all of their physiological systems are geared toward survival in those conditions.

The general rule for growing succulents is to give them as much light as possible in conjunction with good air movement. Grow them in a well-draining soil, and when you water them, water deeply and thoroughly. Let them dry completely and then stay dry for several days before watering again. Succulents need water like any plant; they just need it less often. They also like balanced fertilizer, but again, less often since they are not "programmed" to grow rapidly.

If grown well, many succulents will gift you with surprisingly large and colorful, if fleeting, flowers. We tend to love succulents not for their flowers, though, but for their unusual body designs and spiny or hairy adornments. These unique, bold features are what make succulents as a whole such a rich source of "weird" plants. Walk into a desert display at any botanical garden and you feel like you have a stepped into a book by Dr. Seuss, populated by a menagerie of strange but somehow familiar creatures. Some of them may even look more familiar than we'd care to admit!

Cacti and Euphorbias

We are all guilty of calling distinctly different products by the same name. We might confuse a Polish sausage for a hot dog or ask to pass the potato chips even if they are made of corn. Some of us refer to CDs as records (some of us old folks, that is). This happens all the time with the two largest groups of stem succulents in the world, the cacti and the euphorbias. They often look alike, and many people simply call any globular or tall, spiny, leafless succulent a cactus. It's kind of like calling all facial tissues Kleenex. If plants were like people, euphorbias might develop a complex.

Despite first appearances, these plants do have differences. True cacti are strictly from the New World, have showy flowers, and have spine clusters (called areoles) along their ribs. Cacti rarely have milky sap and are not poisonous. Euphorbias, on the other hand, are mostly African and invariably have a milky sap that can be quite caustic. They have tiny flowers (colorful but usually not flashy) and normally just two prickles at each joint along the ribs, though these are no less dangerous than the spines of cacti. These two groups are the best examples of convergent evolution. From a growing standpoint, they have identical cultural requirements. Who can blame people for lumping them together? Except we all like to be appreciated for our individuality, at least every now and then. I'm proud of my heritage and my family—I'm a euphorbia, darn it!

Old man cacti are covered in white hairs that help keep them cool in the hot sun.
Photo by Larry Mellichamp.

OLD MAN CACTUS
Cephalocereus senilis

PLANT TYPE: Herbaceous perennial
HEIGHT AND SPREAD: 12 in. × 4 in. (30 cm × 10 cm)

YOU JUST WANT TO CARESS these beautiful cacti. Their long hair feels a little rough but silky just the same. Then suddenly, ouch! There are spines under those white threads, and real ones, too. You'd think this cactus was *trying* to get you stuck. The purpose of the hair, however, is not to attract but to reflect. The white hairs reflect sunlight and keep the plants a little cooler in their harsh desert environments. The more hair on a cactus, the more sun it likes—well, actually, the more sun it is trying to protect itself from. The spines are present for the usual reason: to keep the plant from being eaten.

These cacti belong to genera such as *Cleistocactus*, *Mammillaria*, *Neoporteria*, and *Oreocereus*—any cactus with unusually long hair can be called an "old man." The most famous species, however, is *Cephalocereus se-nilis*, whose name means "old head cactus." Ironically, these cacti take an extremely long time to grow large enough to bloom (a plant's sign of maturity), so the "old men" we grow in pots are usually youngsters. Some old man cacti are short and clump-forming, others tall and slender. If you do try to grow one, we recommend providing an "old woman cactus" as a companion. Otherwise they become curmudgeonly, even prickly!

GROWING TIPS

DIFFICULTY RATING: 2
INDOORS VS. OUTDOORS: Indoors but may be grown outdoors in summer
LIGHT: Very bright light
HARDINESS: Not hardy
MOISTURE: Water thoroughly. Let dry completely between waterings. Do not overwater—wait another day or two if unsure. In winter provide even less (practically no) water
GROWING MEDIUM: Good potting mix with 50% perlite or coarse sand
NOTES: Tolerates very high temperatures in summer, down to 40°F (4°C) in winter

Did You Know?
The more hair on a cactus, the more sun it likes—well, actually, the more sun it is trying to protect itself from.

EPIPHYLLUM

Epiphyllum guatemalense var. *monstrose* "Curly Locks"

PLANT TYPE: Herbaceous perennial
HEIGHT AND SPREAD: 12 in. × 12 in. (30 cm × 30 cm) or larger

THIS WEIRD CACTUS MUST BE A CROOK, since it just can't seem to grow straight! It has a genetic deformity that prevents the stems from growing normally. However, it comes true from seed by a strange phenomenon whereby viable seeds actually form without pollination, and so act as asexual propagules. This plant is a bit of an exception to our rule of only presenting "natural" plants that breed true from seed. While it does come out with twisted growth from seeds, those seeds are not the result of normal sexual (mixing up of genes) reproduction. *Epiphyllum guatemalense* var. *monstrose* "Curly Locks" is a true member of the family Cactaceae and comes from Mexico, Guatemala, and Honduras. The variety name *monstrose* means "monstrosity" or "abnormality," clearly referring to the contorted growth form. "Curly Locks" is related to the famous night-blooming cereus, *E. oxypetalum*, which has larger flowers and wider, straight or arching stems. The term "night-blooming cereus" is actually used for as many as six different cacti in different genera.

"Curly Locks" produces long, flat stems (no leaves) 1–3 in. (2.5–8 cm) wide that twist and curl as they grow and are often displayed in a hanging basket. Un-

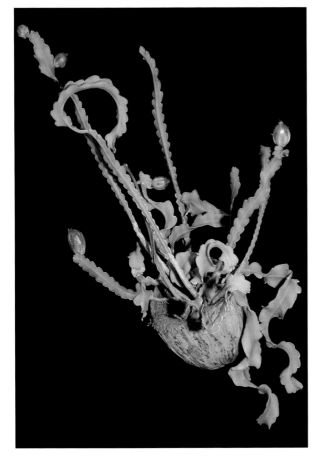

Epiphyllum guatemalense var. *monstrose* "Curly Locks" has contorted stems and pink fruits. Photo by Richard H. Gross.

der good growing conditions the stems may reach 2 ft. (61 cm) long or more. The fragrant flowers open for one night only. They are pure white, about 6–8 in. (15–20 cm) long, with a very slender tube and many narrow petals forming a funnel-shaped flower. After flowering, the ovary may ripen into a pink, fleshy, slightly oblong fruit about 1 in. (2.5 cm) long, containing many black seeds in a pulpy mass. It produces several flowers a year in summer, but don't blink or you'll miss them. The bright pink fruits are long-lasting, though, and provide an additional intrigue to the twisty, twirly green stems of this rogue charmer.

BASEBALL PLANT
Euphorbia obesa

GROWING TIPS

DIFFICULTY RATING: **2**
INDOORS VS. OUTDOORS: Indoors
LIGHT: Very bright light
HARDINESS: Not hardy
MOISTURE: Water thoroughly.
  Let dry completely between
  waterings. Do not overwater—
  wait another day or two if
  unsure. In winter provide even
  less (practically no) water
GROWING MEDIUM: Good potting
  mix with 50% perlite or coarse
  sand

Did You Know?
*Some growers have reported
that one plant will produce
male-only flowers one year and
female-only flowers the next.*

PLANT TYPE: Herbaceous perennial
OTHER COMMON NAME: Golf ball cactus
HEIGHT AND SPREAD: 2 in. × 2 in. (5 cm × 5 cm)

THIS PLANT HAS BEEN REPORTED to change its sex! When a human wants to do this, it requires a medical operation. Plants, on the other hand, are already bisexual by nature, usually producing both male and female parts in the same flower. A significant number of them do have flowers that *express* only

Baseball plant (*Euphorbia obesa*) is about as round as a plant can be, to reduce surface area and conserve water. Photo by Lourdes.

one sex. The missing parts are still "there" but are vestigial and genetically suppressed. So, a plant can have separate functionally male flowers and functionally female flowers. Squash plants are like this, having separate male and female flowers on the same plant. This condition absolutely requires cross-pollination. Other plants have separate male and female flowers on separate plants, but they can't switch back and forth. To flip-flop between sexes on the same individual is very rare in plants, just as in animals, but it can happen. The common eastern North American wildflower known as jack-in-the-pulpit can have male-only flowers when young and switch to female-only flowers as it grows older and stronger. If it is weakened by drought or shade, it can revert to making the less complicated male-only flowers.

Now we come to baseball plant. Though sometimes called "golf ball cactus," it is not really a cactus. It is in the family Euphorbiaceae and comes from the Great Karoo desert of South Africa. Its New World look-alike is the sea urchin cactus, *Astrophytum asterias* (Cactaceae), from the Rio Grande region of Texas and Mexico, which has showy, yellow, bisexual flowers that never change sexes. Both plants are almost perfectly round and lack spines. Their rotund structure produces the smallest surface area relative to their water-storage volume, limiting exposure to the dry air and keeping water loss to a minimum. *Euphorbia obesa* grows slowly and will produce small, nonshowy flowers (either male or female—see also *E. antisyphilitica*, page 132) in spring if it receives enough water and fertilizer at the right time. Some growers have reported that one plant will produce male-only flowers one year and female-only flowers the next. This could be due to maturation, or a response to more nutrients (female flowers need more food to mature the seeds).

Now that you know of the special nature of baseball plant, be kind to your spherical succulent friend. She or he is doubly unique. To be perfectly round is an oddity, but to also change sex is just plain odd.

Caudiciforms

Clearly a title like that calls for an explanation. The term *caudiciform* refers to a growth habit, or form, of particular succulents. Caudiciform succulents have a caudex, an enlarged base that can be formed from either root or stem tissue. Typically the organ is found underground, where it is enlarged to store water and protected from drying sun and wind. The aboveground portion of the plant may be a branch or a twining stem, growing during the rainy season and dying back in the dry season to leave the caudex only. Or the caudex can be the actual lower stem of the plant, or trunk, and be above ground all the time, often grotesquely misshapen by normal standards due to exposure to a harsh desert environment. Even leaves can be modified into water-storing structures in the form of enlarged underground bulbs (think of an onion with its leafy layers), sending up temporary growth during good times.

Some caudiciforms are tall and treelike, but others are low and squat, even creeping. Humans find these forms interesting and grow them in pots with the caudex proudly displayed above ground as a prominent feature. The best caudiciforms must be grown as intact plants from the seedling stage, as the swollen part is usually developed from the hypocotyl, the region between the true root and true stem of the seedling. This means you can't just root cuttings and get the most favorable form of the caudex. So, a well-sized caudiform is treasured by its owner. Some of these can look like (and certainly be valued like) a bonsai specimen. And the more bizarre, the better.

CLIMBING ONION

Bowiea volubilis

PLANT TYPE: Herbaceous perennial
HEIGHT AND SPREAD: 36 in. × 6 in. (91 cm × 15 cm) or
  larger

TRY UNTANGLING THE GROWTH of this unique succulent and stretching it out—you may find it is your longest potted plant. The mass of stringy tissue is huge but temporary.

South African *Bowiea volubilis* (Liliaceae) is a green, brittle, winter-growing vine up to 9 ft. (2.7 m) long. The genus is named after a British plant explorer, and the specific epithet means "twining." In nature it grows from a large underground bulb, but in pots it is usually planted on top of the soil so that you can see the peeling brownish layers that surround each other to form a larger ball. The vine grows out the top of the bulb during the growing season, starting in fall, and dries up to crispy desert debris at the end in late spring. There are no leaves, just green stem tissue that forks and grows rapidly, twining around your other plants unless you give it a separate trellis or support. Late in growth, greenish white flowers about $1/2$ in. (1 cm) wide form on the tips of some of the vine prongs. The sexes are separate, so it takes two different plants to achieve cross-pollination and produce seeds. New plants, however, may be easily grown from little asexual bulblets that form around the edges of the older bulb scales.

Give this plant lots of light and plenty of water

GROWING TIPS

DIFFICULTY RATING: **2**
INDOORS VS. OUTDOORS: Indoors
LIGHT: Very bright light
HARDINESS: Not hardy
MOISTURE: Keep moist during
  growth but dry in summer
GROWING MEDIUM: Good potting
  mix with 50% perlite or coarse
  sand
NOTES: Needs a trellis for vine
  support

Did You Know?

There are no leaves, just green stem tissue that forks and grows rapidly, twining around your other plants unless you give it a separate support.

Climbing onion (*Bowiea volubilis*) produces a mass of twining stems from its onionlike bulb. Photo by Larry Mellichamp.

while in growth, but keep it dry in summer. In time a clump of large bulbs will form, and you will see their resemblance to grocery store onions—at least until that Silly String mass of green stems emerges. Getting to appreciate this yearly cycle is part of what makes climbing onion a joy to grow. Just when you start eyeballing those bulbs for onion rings, it reminds you that it is no ordinary onion and that you had better find something for it to climb on if you want to see it standing up for the next five months!

Euphorbia francoisii

PLANT TYPE: Herbaceous perennial
HEIGHT AND SPREAD: 4 in. × 12 in. (10 cm × 30 cm)

EUPHORBIA FRANCOISII HAS NO COMMON NAME. It comes from Madagascar, where it is very rare and perhaps endangered from habitat destruction by the locals. It does not seem to have local uses, appeal, or friends—hence, no local name. A botanist named it after a local landowner, Mr. Francois, although the pronunciation of the specific epithet seems to vary. (Latin names promise one spelling but are still up for interpretation in the speaking of them.)

No matter what you call it, this is an interesting caudiciform. Its thick, water-storing root branches to form aboveground stems with leaves, and the crinkled leaves and dull flowers are appealing as a contrast to the contorted caudex. *Euphorbia francoisii* is easily grown, and no two plants have exactly the same beautiful leaf markings.

We don't need a common name to grow and enjoy *Euphorbia francoisii*. But when we try to tell our friends how great it is, using that Latin name sometimes makes us think about where it came from back in Madagascar and why no one seemed to care enough to give it a local name.

Euphorbia francoisii has a massive underground root caudex and branching leafy stems that we like to admire by growing high in a pot. Photo by Paula Gross.

Stapeliads

We stare as we drive by—a dead animal on the roadside, rotting in the sun, it hair coated in dried blood and blowing in the breeze, perhaps covered in green blowflies laying their eggs in the stinking flesh. Not a pretty sight to us humans (and yet we must look!). So why would we want to grow plants that look and smell like dead animals and attract flies for pollinators? Because they're odd, interesting, and even eerily beautiful.

This group of African stem succulents is called stapeliads because they have a lot in common with the large and well-known genus *Stapelia*. All are desert-dwelling and produce thickened, angled stems. They only require sparse watering in summer and even less in winter, and will rot easily if overwatered. They make large colonies or clumps and can actually creep along the ground. But it's their flowers where the real interest lies.

Starfish flower (*Stapelia hirsuta*) produces a hairy flower that looks and smells like roadkill. Photo by Larry Mellichamp.

Stapeliads are members of the milkweed family (Asclepiadaceae) and share certain flower structure similarities with the Northern Hemisphere butterfly weeds in the genus *Asclepias*. The flowers are star-shaped, hence a common name of starfish flower, and come in various shades of red, maroon, black, brown, and yellow—all the colors of old, bloody meat. Many species also have hairs on their flowers to mimic dead animal fur. Then there's the smell—the awful smell, the nauseating smell, the smell that only a fly can love. And that's how they trick carrion flies to act as pollinators. The flies come to the flowers, think they are rotting flesh, lay their eggs, and leave, carrying pollen to another flower. The eggs hatch, but the maggots have nothing of substance to feed on, so they die. If this were to happen too often, the fly population would plummet, but these plants only bloom sporadically in nature, and there are still plenty of real animal corpses in the desert to feed the flies.

One common stapeliad is the starfish flower, *Stapelia hirsuta*, whose specific epithet means "hairy." Another one, *S. gigantea*, has creeping stems as thick as a broom handle and flowers more than 10 in. (25 cm) across. *Stapelia gigantea* is easy to grow and not difficult to propagate, so folks often share it with one another. But in late summer, when it blooms and is indoors, you may wonder about your friend's generosity. I actually knew a grown man who took one and buried it in the backyard in disgust. The lifesaver plant, *Huernia zebrina*, whose name means "striped," has a glossy, colorful, almost whimsical appeal. The top of the flower has a swollen ring that looks like its common name. It has a weak smell and is compact and free-flowering—a good one to grow. *Caralluma speciosa*, another striking stapeliad, grows on very thick, four-angled, upright stems. It becomes crowned in late summer by a ball

A *Stapelia* flower has attracted green bottle flies who have laid their eggs and may act as pollinators. Photo by Martin Heigan.

Lifesaver plant (*Huernia zebrina*) looks like it has a colorful inner tube. Photo by Larry Mellichamp.

of 1 in. (2.5 cm), black, yellow-centered flowers that have little black hairs that wave along the petal edges. And boy, does this plant smell bad! The saving grace is that the flowers last only a few days. Still, I look forward to stapeliad flowers every year—a ritual of disgust that attracts one of the largest followings of any succulent plant group.

Mimicry Succulents

If a human tried to live in the desert without water, he or she would be dead in a week, yet some plants can live in deserts that get virtually no predictable rainfall. How do they do it? They would certainly have to be able to store and retain water very efficiently. They wouldn't grow very much. They'd also need protection to keep from being eaten by the first thirsty animal that came along. But why would anything want to grow in a place like that anyway? We cannot answer that question other than to say that plants (and animals) do what they can, and live wherever they can, because they can. They have evolved the special adaptations that allow them to survive and reproduce in such harsh places, where there is minimal competition. We cannot ask why, only how.

Mimicry succulents protect themselves by "hiding"—blending in, being overlooked. Most of them are almost round and look like stones. In fact they are called living stones, and the genus they belong to, *Lithops* (Mesembryanthemaceae), means "stonelike." Besides their size and shape, they often have color patterns that make them look even more like colorful desert pebbles. Since they have no direct protection from being eaten, they have to rely on camouflage. Being so small and short, they can be mostly covered with sand such that the tips of the rounded leaves just barely peek out. And speaking of leaves, the plants are almost all leaf—a pair of leaves that are quite plump with water, connected to a tiny taproot and stem buried underground. They live a life of minimal exposure, which keeps them from experiencing the full brunt of the desert heat. The water they do get, an average of just a few inches of rainfall per year, is quickly absorbed and stored. Mimicry succulents also can get moisture from the heavy daily fogs that form in some of the driest parts of western South Africa and Namibia where they live. They can shrivel quite a bit before being revived with the rains or fogs. In autumn, just before going dormant, they produce beautiful flowers that are white or yellow and not at all camouflaged (to attract pollinators). Seeds are produced that are kept inside ornate seedpods until a rain occurs of sufficient magnitude to allow for ger-

GROWING TIPS

DIFFICULTY RATING: 2

INDOORS VS. OUTDOORS: Indoors

LIGHT: Full sun

HARDINESS: Not hardy

MOISTURE: Water thoroughly as needed during summer growing season, but let dry completely between waterings

GROWING MEDIUM: Good, well-draining potting mix with 50% sand or perlite added

NOTES: Mimicry succulents rot readily if kept too wet. Give good air movement. During winter dormancy, water just enough to keep from shriveling, perhaps once a month. Fertilize sparingly

Each living stone plant (*Lithops*) consists of a pair of leaves with a slit in between. It sheds its old leaf skin as it grows a new set of leaves each year when the rains come. Photo by Larry Mellichamp.

mination and establishment; then the seedpods open, and the seeds splash out by raindrop action. The next generation of harsh desert-dwellers is born and will hopefully find a suitable spot to live and hide.

Window-Leaved Succulents

Here's a paradox: Windows can be a *pane*. They let in light and provide a barrier from outside temperatures. Windows can also be a *pain*. They let in light, but some of this light converts to heat and can build up inside the room, resulting in a greenhouse effect. In the desert, sunlight is your friend and your enemy, because it is so abundant and intense. It is required for photosynthesis but can result in overheating of the leaf. What to do?

Window-leaved succulents have found an adaptation to the harshness of desert light. Their highly succulent leaves (with much reduced stems) live under the sand, where they stay relatively cool even when the temperature on the surface reaches 110°F (43°C). They get their light through tiny "windows"—the tips of their succulent leaves, which are translucent. These leaves grow so that they just peek through the sand and see a little sun. The energy-rich rays enter through the window and bounce around inside the central core of the thick leaf, which is a spongy mass of clear cells. This allows photosynthesis to take place in the outer cortex, which is green with chlorophyll. Like looking through a pinhole at a solar eclipse, you let in enough light but keep out most of it to avoid damage.

Several different families include many different window-leaved succulents. *Haworthia cooperi*, in the Liliaceae, comes from South Africa and has fat, little, angular leaves whose windows have regions of translucent pigment. Baby toes, *Fenestraria rhopalophylla* (Mesembryanthemaceae), with its large white flower,

GROWING TIPS

DIFFICULTY RATING: **2**
INDOORS VS. OUTDOORS: Indoors
LIGHT: Full sun
HARDINESS: Not hardy
MOISTURE: Water thoroughly as needed during summer growing season, but let dry completely between waterings
GROWING MEDIUM: Good, well-draining potting mix with 50% sand or perlite added
NOTES: Window-leaved succulents rot readily if kept too wet. Give good air movement. During winter dormancy, water just enough to keep from shriveling, perhaps once a month. Fertilize sparingly

A window-leaved plant, *Haworthia cooperi*, has plump leaves with translucent areas at the tips to allow a little sunlight to penetrate down inside. Photo by Larry Mellichamp.

The leaves of baby toes (*Fenestraria rhopalophylla*) are covered by sand with just the tips exposed so that light can enter through the triangular window tip. The flower emerges from the sand for pollination. Photo by Larry Mellichamp.

is one of the most beautiful window-leaved succulents. The tips of its tubular leaves have a triangular window, while the rest of the leaf is under the sand. Among the other fascinating species with window leaves are green-pea vine (*Senecio rowleyanus*, Asteraceae), erect peperomia (*Peperomia columnaris*, Piperaceae), other *Haworthia* species, and several species of living stones, including *Lithops* and *Conophytum*. Each provides its own little window into the wonderful world of succulents.

We grow these plants in a pot so that we can see their beautiful leaf textures and windows. In bright light they often take on a tinge of pink because of a red pigment that is produced to help protect them from the sun's rays. If it were really hot, the plants would burn from overexposure to direct sun—just like white folks who go to the beach in spring, suddenly exposing their sensitive skin to sun, and instantly burn.

Leaf Succulents

In the desert, leaves can be a liability. They're thin, lose water, and dry up. They're also easy for an herbivore to eat. Normally they have to drop off during the dry, dormant season and be regrown. But many desert plants have evolved succulent water-storing leaves, as opposed to succulent stems or roots. The mimicry and window-leaved plants mentioned above consist of mostly leaf tissue. Leaves are the "normal" organs for photosynthesis, and many desert plants evolved to retain them for this purpose, while adding the water-storing cells that make them thick and juicy. Sometimes they are born on the elongated stems of a shrublike plant (for example, the well-known jade tree). Often they can be on short stems and form a rosette—a tight cluster of leaves arranged in a series of rings or spirals, like a lettuce head. In any case, they need some protection, and this usually comes in the form of acids or poisonous saps. The rosette leaf succulents can be quite attractive in the patterns they form with overlapping leaves. One of the most spectacular, but one that we don't feature here because it is unavailable and difficult to grow, is the spiral aloe, *Aloe polyphylla*. Here are just a few growable leaf succulents that strike our fancy.

BLACK TREE AEONIUM
Aeonium 'Zwartkop'

PLANT TYPE: Herbaceous perennial
HEIGHT AND SPREAD: 12 in. × 12 in. (30 cm × 30 cm)

WE COULD HAVE FEATURED a whole group of plants with black foliage or flowers in this book, but most don't come true from seed. This one doesn't either, but I just had to sneak it in!

Aeoniums in general are interesting for their tight rosettes of leaves. Tree aeoniums bear these rosettes at the tips of rounded stems and remind me of something you might expect to see growing from a coral reef. Add the black color to the leaves and you really have something weird. Plus it is fun to ask for *Aeonium* 'Zwartkop' at your local greenhouse. The cultivar name is Dutch for "black head," which alludes to the fact that those black (or dark, dark burgundy) rosettes sit perched upon tall, narrow, gray-brown stalks (the necks or bodies on which the "heads" sit). The plant is often multistalked, with those stalks gently curving or crooking, such that the whole affair ends up looking something like a bouquet of black, Wes Craven–inspired flowers. Odd, quirky, and yet actually beautiful.

I envy those in arid zone 10 who can grow this plant outside in a garden. I've seen its powers in pho-

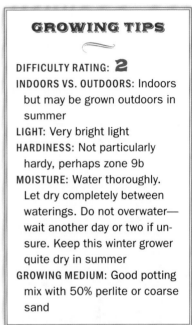

Did You Know?

The plant is often multi-stalked, with those stalks gently curving or crooking, such that the whole affair ends up looking something like a bouquet of black, Wes Craven–inspired flowers.

Black tree aenonium (*Aeonium* 'Zwartkop') has tight rosettes of very dark purple leaves on a branching stem. Photo by Paula Gross.

tos of such gardens, drawing attention to itself while at the same time making its properly chosen neighbors all the more alluring. It is definitely worth growing in a pot for us temperate zoners, though. The leaves may look like black flowers, but they have nothing to do with sexual reproduction. The plant has true flowers, of course, but it is not particularly anxious to bloom. This is actually a good thing if you are growing it, for each stalk will die after setting seed. Plants such as this are referred to as monocarpic (living several years, then flowering once and dying), and most aeoniums behave this way. If it's going to happen, though, enjoy it. The panicles of yellow flowers are amazing set against that dark foliage. Now that's "going out with a bang."

MOUSTACHE PLANT
Aloe suprafoliata

PLANT TYPE: Herbaceous perennial
HEIGHT AND SPREAD: 12 in. × 24 in. (30 cm × 61 cm)

AREN'T BABIES CUTE, with their fat little cheeks and big grins showing their baby teeth? They may have lots of hair, too—or none. But babies don't stay small forever, and it's not too long before their hormones kick in and they change into their not-quite-so-cute preteenage form, followed by even more ungainly rites of puberty. Plants go through juvenile changes as well. Sometimes these changes are barely noticeable, but in some plants they are wildly apparent.

Moustache plant, *Aloe suprafoliata* (Liliaceae), is a winter-growing South African plant that has dual growth forms. When the plant is a youngster, the first one to four years of its life, the rosette (leaf cluster) forms gray-green, spiny-toothed leaves that lie directly nestled on top of one another in one plane to the left and right. *Suprafoliata* means "with superimposed leaves." At this stage the two growing sides of the plant seen together from above or from the side look like a symmetrical, albeit blue-green, moustache. As the plant matures, hormones kick in, and suddenly one autumn the newly forming leaves begin to develop in a spiral from the center of the plant, and get much larger—more than 1 ft. (30 cm) long. The adult plant proceeds to develop into a rosette over 2 ft. (61 cm) across, with very thick, very spiny leaves. Not so cute anymore. In fact it's a bit vicious. Protec-

GROWING TIPS

DIFFICULTY RATING: **2**
INDOORS VS. OUTDOORS: Indoors but may be grown outdoors in summer
LIGHT: Very bright light
HARDINESS: Not hardy
MOISTURE: Water thoroughly. Let dry completely between waterings from September to April. Do not overwater—wait another day or two if unsure. Keep this winter grower quite dry in summer
GROWING MEDIUM: Good potting mix with 50% perlite or coarse sand

tion is the name of the game out there in the wild desert, especially when you're bearing flowers full of nectar and then precious fruits containing the seeds of your next generation. When grown in bright light to full sun, the adult plant can make several tall, unbranched stalks producing numerous 2 in. (5 cm), red-orange, tubular flowers. The flowers are beautiful and worth waiting for.

As with most aloes, moustache plant requires two genetically different plants to cross-pollinate and get seeds. If you want to always have young plants with their nifty foliage formation, get two adults, cross-pollinate them to get seeds, and grow new plants. On the other hand, young plants sprout readily from the base of the adult, and these sprouts can be potted up and grown into nice little babies that will, for a few years, exhibit the overlapping leaf phenomenon—but they will not be genetically distinct from each other. If you carefully pry them away from the adult with roots intact, you will have bleeding hands and young plants to boot. No one ever said it was painless to raise children, no matter how cute they can be.

Moustache plant (*Aloe suprafoliata*) is a juvenile form with strictly overlapping leaves. Photo by Larry Mellichamp.

MOTHER OF THOUSANDS
Kalanchoe delagoensis

PLANT TYPE: Herbaceous annual or perennial
HEIGHT AND SPREAD: 12 in. × 6 in. (30 cm × 15 cm) or
   larger

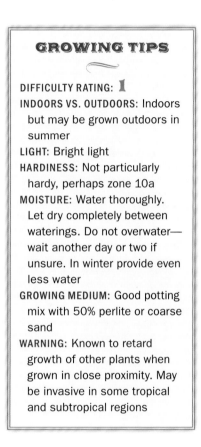

GROWING TIPS

DIFFICULTY RATING: **1**
INDOORS VS. OUTDOORS: Indoors
  but may be grown outdoors in
  summer
LIGHT: Bright light
HARDINESS: Not particularly
  hardy, perhaps zone 10a
MOISTURE: Water thoroughly.
  Let dry completely between
  waterings. Do not overwater—
  wait another day or two if
  unsure. In winter provide even
  less water
GROWING MEDIUM: Good potting
  mix with 50% perlite or coarse
  sand
WARNING: Known to retard
  growth of other plants when
  grown in close proximity. May
  be invasive in some tropical
  and subtropical regions

MOTHER OF THOUSANDS is one of those plants with a really cool feature that initially charms. In this case it is the multitude of little baby plantlets, perfectly formed and arranged along the margins of the leaves. How cute! But the more you learn about the plant, the more you get a sinking feeling that this "mom" is not so wholesome and caring and may be out to nurture nothing but itself.

From its own point of view, *Kalanchoe delagoensis* (syn. *Bryophyllum tubiflorum*) is one successful mother. After all, it can produce plantlets and propagate itself without the need for sexual reproduction. Never mind the fact that it has become an invasive weed in certain areas of the world or that it is deadly poisonous to livestock. Heck, these are just survival strategies for this drought-tolerating succulent, right? Still, not so cute.

The poison in this plant and some of its relatives belongs to a class called bufadienolides. These lovely little compounds are cardiac poisons, can cause cell growth inhibition, and sometimes cause nerve and muscular system disruptions. Mother of thousands is native to Madagascar but has become a naturalized weed in South Africa, where its ingestion is blamed for the death of more than 12,000 cattle a

Mother of thousands (*Kalanchoe delagoensis*) produces
countless offspring from along the edges of the leaves.
Photo by Larry Mellichamp.

year. This represents not a nuisance but a serious problem for cattle farmers, not to mention cows. A better name for this plant in southern Africa might be "killer of thousands."

Did I mention that it is strikingly architectural and handsome when
it blooms? That it is super easy to grow with reasonable light and thrives
on neglect? Thought I'd better bring this back around from death and
doom. The plant is intriguing and you may want to grow it, but just know
what you are signing on for. Don't grow it outside south of zone 9, and
keep animals and kids away. Grow it inside in a pot by itself, give it light,
and neglect it. You will see just how determined this supermom is to live
and "give birth" to a multitude of clonal babies.

SPEAR SANSEVIERIA
Sansevieria cylindrica

Did You Know?
Making the leaves cylindrical, rather than flattened, reduces their surface-to-volume ratio, which means less surface from which to potentially lose water.

PLANT TYPE: Herbaceous perennial
HEIGHT AND SPREAD: 12 in. × 6 in. (30 cm × 15 cm) or larger

IF I AM ASKED WHAT'S SO GREAT about spear sansevieria and I simply answer, "It has round leaves," I will be met with one of those looks that says, "Are you sure you work here?" Lots of plants have round leaves: geraniums, water lilies, nasturtiums. Maybe I need to try again. How about not round leaves, but rather leaves in the round—cylindrical leaves, like striated green broom handles sticking up out of the ground. If nature had its own Ikea, the first thing on the plant aisle would be *Sansevieria cylindrica* (formerly in the Liliaceae, now the Ruscaceae). This plant is a modern minimalist-style decorator's fantasy. All those frills and curls of most plants must frustrate Swedish designers terribly.

Of course, nature didn't design this plant to appeal to our twenty-first-century styling. The environment has shaped it to be wonderfully adapted to intense sunlight and very little moisture. Making the leaves cylindrical, rather than flattened, reduces their surface-to-volume ratio, which means less surface from which to potentially lose water. Tall, thin, spearlike leaves growing straight up don't give the hottest overhead sun of the day anything to shine on directly. And anything that keeps the desert sun from baking your tissues is a good thing. The actual stem

The leaves of *Sansevieria cylindrica* are cylindrical and erect to avoid direct hot sun. Darker banding provides camouflage. Photo by Larry Mellichamp.

of spear sansevieria avoids the sun altogether, growing below the soil surface as a rhizome.

This Angolan species joins about seventy other *Sansevieria* species from the African and Asian deserts and tropics. The most familiar is *S. trifasciata* and its many cultivars, which carry the slightly threatening common names of snake plant, devil's tongue, or mother-in-law's tongue (ouch). *Sansevieria cylindrica* is extremely neglect-proof (as is *S. trifasciata*)—a perfectly modern-styled plant for a minimalist-styled caretaker.

Hardiness Zone Temperatures

Temperatures

$°C = 5/9 \times (°F-32)$
$°F = (9/5 \times °C) + 32$

Plant Hardiness Zones

Average Annual Minimum Temperature

ZONE	TEMPERATURE (DEG. F)			TEMPERATURE (DEG. C)		
1	below		−50	−45.6	and	below
2a	−45	to	−50	−42.8	to	−45.5
2b	−40	to	−45	−40.0	to	−42.7
3a	−35	to	−40	−37.3	to	−40.0
3b	−30	to	−35	−34.5	to	−37.2
4a	−25	to	−30	−31.7	to	−34.4
4b	−20	to	−25	−28.9	to	−31.6
5a	−15	to	−20	−26.2	to	−28.8
5b	−10	to	−15	−23.4	to	−26.1
6a	−5	to	−10	−20.6	to	−23.3
6b	0	to	−5	−17.8	to	−20.5
7a	5	to	0	−15.0	to	−17.7
7b	10	to	5	−12.3	to	−15.0
8a	15	to	10	−9.5	to	−12.2
8b	20	to	15	−6.7	to	−9.4
9a	25	to	20	−3.9	to	−6.6
9b	30	to	25	−1.2	to	−3.8
10a	35	to	30	1.6	to	−1.1
10b	40	to	35	4.4	to	1.7
11	40	and	above	4.5	and	above

To see the U.S. Department of Agriculture Hardiness Zone Map, go to the U.S. National Arboretum site at http://www.usna.usda.gov/Hardzone/ushzmap.html.

BIBLIOGRAPHY

Allan, Mea. 1977. *Darwin and His Flowers: The Key to Natural Selection*. New York: Taplinger.

Attenborough, David. 1994. *Private Life of Plants*. London: BBC.

Barlow, Connie. 2000. *Ghosts of Evolution: Nonsensical Fruits, Missing Partners, and Other Ecological Anachronisms*. New York: Basic Books.

Barthlott, Wilhelm, Stefan Porembski, Rüdiger Seine, and Inge Theisen. 2007. *The Curious World of Carnivorous Plants: A Comprehensive Guide to Their Biology and Cultivation*. Portland, Oregon: Timber Press.

Behme, Robert Lee. 1992. *Incredible Plants: Oddities, Curiosities and Eccentricities*. New York: Sterling.

Bender, Steve, and Felder Rushing. 1993. *Passalong Plants*. Chapel Hill, North Carolina: University of North Carolina Press.

Berg, Linda. 2007. *Introductory Botany: Plants, People, and the Environment*. Fort Worth, Texas: Saunders. (A very good basic botany textbook.)

Bernhardt, Peter. 1990. *Wily Violets and Underground Orchids: Revelations of a Botanist*. New York: Vintage Books.

Bernhardt, Peter. 1993. *Natural Affairs: A Botanist Looks at the Attachments Between Plants and People*. New York: Villard Books.

D'Amato, Peter. 1998. *The Savage Garden: Cultivating Carnivorous Plants*. Berkeley, California: Ten Speed Press.

Dressler, Robert L. 1981. *The Orchids: Natural History and Classification*. Cambridge, Massachusetts: Harvard University Press.

Elbert, Virginie F., and George A. Elbert. 1975. *Fun with Growing Odd and Curious House Plants.* New York: Crown.

Emboden, William A. 1974. *Bizarre Plants: Magical, Monstrous, Mythical.* New York: Macmillan.

Graham, Pamela. 2000. *Special and Strange! Unusual Plants.* Kew, Australia: Troll Communications. (Especially for children.)

Gusman, Guy, and Liliane Gusman. 2006. *The Genus* Arisaema: *A Monograph for Botanists and Nature Lovers.* Ruggell, Lichtenstein: A .R. G. Gantner Verlag.

Hoshizaki, Barbara Joe, and Robbin C. Moran. 2001. *Fern Grower's Manual.* Revised and expanded edition. Portland, Oregon: Timber Press.

Loewer, Peter. 1993. *The Evening Garden.* New York: Macmillan.

Loewer, Peter, and Larry Mellichamp. 1997. *The Winter Garden: Planning and Planting for the Southeast.* Mechanicsburg, Pennsylvania: Stackpole Books.

Mann, John. 1992. *Murder, Magic, and Medicine.* Oxford, England: Oxford University Press.

Meeuse, Bastiaan, and Sean Morris. 1984. *The Sex Life of Flowers.* New York: Facts on File.

Mellichamp, Lawrence. 1994. Are you stuck on the fine points of sharp-object nomenclature? *Cactus and Succulent Journal* 66: 208–213.

Mellichamp, Lawrence. 2008. The *Sarracenia* pitcher plants and bog gardening. *Sibbaldia* 6: 79–99.

Menninger, Edwin A. 1967. *Fantastic Trees.* New York: Viking.

Mulligan, William C., and Elvin McDoland. 1992. *The Adventurous Gardener's Sourcebook of Rare and Unusual Plants.* New York: Simon and Schuster.

Orlean, Susan. 1998. *The Orchid Thief: A True Story of Beauty and Obsession.* New York: Random House.

Overy, Angela. 1997. *Sex in Your Garden.* Golden, Colorado: Fulcrum.

Preston-Mafham, Rod, and Ken Preston-Mafham. 1991. *Cacti: The Illustrated Dictionary.* London: Blandford.

Pollan, Michael. 2001. *The Botany of Desire: A Plant's-Eye View of the World.* New York: Random House.

Proctor, Michael, Peter Yeo, and Andrew Lack. 1996. *The Natural History of Pollination*. Portland, Oregon: Timber Press.

Rice, Barry A. 2006. *Growing Carnivorous Plants*. Portland, Oregon: Timber Press.

Rowley, Gordon. 1978. *The Illustrated Encyclopedia of Succulents*. New York: Crown.

Sajeva, Maurizio, and Mariangela Costanzo. 1994. *Succulents: The Illustrated Dictionary*. Portland, Oregon: Timber Press.

Schnell, Donald E. 2002. *Carnivorous Plants of the United States and Canada*. Portland, Oregon: Timber Press.

Talalaj S., D. Talalaj, and J. Talalaj. 1991. *The Strangest Plants in the World*. Melbourne: Hill of Content.

Temple, Paul. 1985. *How to Grow Weird and Wonderful Plants*. London: Beaver Books. (Especially for children.)

Wilkins, Malcolm B. 1988. *Plant Watching: How Plants Remember, Tell Time, Form Partnerships and More*. New York: Macmillan.

INDEX

Aeonium 'Zwartkop', 264–265

Aeschynanthus radicans, 183

air plant, 163–165

Albizia julibrissin, 162

Aloe polyphylla, 263

Aloe suprafoliata, 267–268

Amaranthaceae, 172

Amaranthus caudatus, 119–121

Amorphophallus bulbifer, 135–137

Amorphophallus konjac, 137

Amorphophallus rivieri, 137

Amorphophallus titanum, 15, 87–90

Anabaena azollae, 71

Angiopteris evecta, 65

Angraecum sesquipedale, 213–214

Ansellia, 21

Anthurium andraeanum, 92

Anthurium scherzerianum, 91–92

Apiaceae, 96

Araceae, 92, 139

Araucaria araucana, 190, 193–195

Arisaema ringens, 140

Arisaema sikokianum, 140

Arisaema thunbergii, 138–140

Arisaema triphyllum, 139

Aristolochiaceae, 93, 169

Aristolochia grandiflora, 93–95

artillery plant, 145–147

Asarum arifolium, 169–171

Asarum maximum, 169–170

Asclepiadaceae, 129, 257

Asclepias physocarpa, 187

Asplenium bulbiferum, 67–68

Astrophytum asterias, 249

Australian Pitcher Plant, 38–40

Azolla caroliniana, 69–71

Azolla pinnata, 69

baby toes, 261–262

banana, 103–105

baseball plant, 248–249

bat-faced cuphea, 176–177

bat plant, 113–115, 167

bed of nails, 207

beehive ginger, 148–149

Begonia hispida var. *cuculifera*, 155–156

Begonia maculata var. *wightii*, 153–154

Begonia ×*rex-cultorum*, 153

bird-of-paradise, 188–189

birthwort, 93–95

bittersweet vine, 127

black bat plant, 167

black jack-in-the-pulpit, 138–140

black orchid, 229–231

black tree aeonium, 265

bladderwort, 32

bleeding heart, 125–126

blue oil fern, 72–73

Bowiea volubilis, 251–253

bracts, 26, 85, 89, 97, 103, 108

Brassia, 215–217

Brazilian candles, 108–109

Brazilian edelweiss, 24

bromeliad, 163

Bryophyllum tubiflorum. See *Kalanchoe delagoensis*

bucket orchid, 16, 226–228

Bulbophyllum medusae, 218–219

Bulbophyllum phalaenopsis, 222

Bulbophyllum spiesii, 220–221

butcher's-broom, 23, 203–205

butterfly orchid, 236–237

butterwort, 57–58

Cactaceae, 242–243

Calathea, 151

cape sundew, 50

Capsicum annuum 'Peter Pepper', 187

Caralluma speciosa, 257

Cardiospermum halicacabum, 122–123

carnivorous plants, 31

Carolina mosquito fern, 69

Catasetum saccatum, 223–225

Cattleya, 212

caudiciform succulents, 250

Celastrus orbiculatus, 128

Celastrus scandens, 127

Celosia cristata, 172–173

Cephalocereus senilis, 245

Cephalotus follicularis, 38–40

Ceropegia woodii, 129–131

Christmas star orchid, 213–214

cladophyll, 203

Cleistocactus, 245

Cleistocactus winteri, 174–175

climbing onion, 251–253

Clowesia, 231

clubmoss, 76, 79–81

Coal Age, 65, 76, 78

cobra-lily, 9, 41–43

cockscomb, 172–173

coevolution, 214

Congo cockatoo, 180

Conophytum, 262

convergent evolution, 242

corpse flower, 87, 137

Coryanthes macrantha, 16, 226–228

cow's udder, 186

Crassula rhodogyna, 241

Crassulacean acid metabolism (CAM) photosynthesis, 240

cuckoo flower, 180–181

Cuphea llavea, 176–177

curly locks epiphyllum, 246

cyathium, 134, 143

Cymbidium, 212

dancing plant, 157

Darlingtonia californica, 41–43

Darwin, Charles, 32, 45

desert tortoise plant, 22–23

Desmodium gyrans, 157–159

devil's claw, 184–185

devil's thorn, 206–207

devil's tongue, 135, 273

Dicentra cucullaria, 126

Dicentra eximia, 126

Dicentra spectabilis, 125–126

Dionaea muscipula, 44–47

Dioscorea elephantipes, 22–23

Doctrine of Signatures, 95, 118, 126

dodo, 197

Dorstenia yambuyaensis, 141

dove orchid, 234–235

Dracula, 209

Dracunculus vulgaris, 137

dragon arum, 137

Drosera, 48–50

Drosera binata var. *multifida*, 34, 50

Drosera capensis, 50

Drosera capillaris, 49

Drosera glanduligera, 49

Drosera peltata, 50

Drosera regia, 50

Dutchman's-britches, 126

Dutchman's-pipe, 93–95

Epiphyllum guatemalense var. *monstrose* "Curly Locks", 246–247

Epiphyllum oxypetalum, 246

epiphyte, 21, 72, 163, 210

Equisetum, 77–78

Eryngium, 96–97

euglossine bee, 223, 226, 233

Euonymus americanus, 127–128

Euphorbia, 134, 143–144

Euphorbia antisyphilitica, 133, 249

Euphorbia francoisii, 254–255

Euphorbia marginata, 144

Euphorbia obesa, 248–249

Euphorbia pulcherrima, 144

Euphorbiaceae, 143, 249

Fabaceae, 157, 161, 198

family jewels plant, 187

family names, 28

fasciation, 172

Fenestraria rhopalophylla, 261–262

fern allies, 76–83

ferns, 64–76

field horsetail, 78

firethorn nightshade, 206

flamingo flower, 92

floral spur, 180

flower structure, 25

fly pollination, 93–95, 131, 140, 143, 257

flying dragon, 199–200

fork-leaved sundew, 34

fox face, 186–187

Fredclarkeara After Dark 'Black Diamond', 229–231

Gesneriaceae, 182

Gleditsia triacanthos, 196–197

gloriosa lily, 98–99

Gloriosa superba, 98–99

golden rat tail cactus, 175

goldfish plant, 182–183

golf ball cactus, 248

Gomphocarpus physocarpus. See *Asclepias physocarpa*

Gongora, 232–233

Green Swamp Preserve, 36

ground pine, 79

hand plant, 141–142

hanging lobster claw, 85

hawk moth, 218

Haworthia cooperi, 261

hearts-a-burstin', 127–128

heartseed, 123

heart-shaped, 117

Heliamphora, 51–53

Heliconia rostrata, 85, 102

Heliconia stricta, 100–101

Heliconiaceae, 101

Hexastylis. See *Asarum arifolium*

hibernacula, 50, 58

Hildewintera aureispina. See *Cleistocactus winteri*

Holy Ghost orchid, 234

Homalocladium platycladum, 178–179

honey locust, 197

horsetail, 77–78

Hoya carnosa, 25–26

Hoya kerrii, 116—117, 129

Huernia zebrina, 257–258

Impatiens niamniamensis, 180–181

Impatiens psittacina, 181

inflorescence, 133

jack-in-the-pulpit, 138–140

Japanese cobra plant, 140

Japanese fiber banana, 105

Kalanchoe delagoensis, 269–271

labellum, 211

latex, 144

leaf succulents, 263

lifesaver plant, 258

Liliaceae, 99, 251, 261, 267, 272

Linnaeus, Carl, 27

lipstick plant, 182

Lithops, 259–260

little brown jugs, 171

living stones, 259–260

lobster claw, 100–102

love-in-a-puff, 122–124

love-lies-bleeding, 119–121

Lycopodium, 79–81

Lycopodium clavatum, 81

Lycopodium squarrosum, 76

Mammillaria, 17, 245

marsh pitcher plant, 51-53

maypop, 107

medusa orchid, 218–219

Mesembryanthemaceae, 261

Mexican yam, 23

Microsorum thailandicum, 72–73

mimicry succulent, 258–260

Mimosa pudica, 160–162

monkey-cups, 54

monkey puzzle tree, 190, 193–195

Mormodes, 231

mosquito fern, 69–71

moth orchid, 212

mother fern, 67–68

mother-in-law's tongue, 273

mother of thousands, 269–271

moustache plant, 267–268

mule's-foot fern, 65

Musa basjoo, 105

Musa velutina, 105

naranjilla, 207

needle palm, 192

Nematanthus, 182–183

Neoporteria, 245

Nepenthes, 54–56

night-blooming cereus, 246

nightshades, 187

nipple fruit, 186

North American pitcher plant, 59–63

Okefenokee Swamp, 35

old man cactus, 244–245

Oncidium papilio. See *Psychopsis papilio*

Orchidaceae, 210

Oreocereus, 245

oriental bittersweet, 128

ouabain, 111

panda ginger, 169–170

parrot impatiens, 180

Passiflora, 106–107

Passiflora incarnata, 107

passionflower, 106–107

Pavonia (×*gledhillii*, intermedia, multiflora), 108–109

peacock plant, 151

Peperomia columnaris, 262

Peristeria elata, 234–235

Phalaenopsis, 212

pig's ears, 186

piggies ginger, 169–171

piggyback begonia, 156

pigtail plant, 91–92

Pilea semidentata, 145–147

Pinguicula ehlersiae × *P. moranensis*, 57–58

pink banana, 105

pipevine, 93

pitcher plant, 39, 51, 55, 59

   purple, 61

   yellow, 33

   white-top, 61

Platycerium bifurcatum, 76

Platycerium superbum, 74–76

poison arrow vine, 111

polka-dot begonia, 154

pollination, 26

pollinium, 211, 223-225

Poncirus trifoliata "Flying Dragon", 192, 199–200

Proboscidea louisianica, 184–185

pseudobulb, 221

pseudovivipary, 67

Psilotum nudum, 82–83

Psychopsis papilio, 236–237

pyrotechnic spores, 79–81

rattlesnake master, 97

rex begonia, 155

Rhapidophyllum hystrix, 192

ribbon plant, 178

rock tassel-fern, 76

Rosa omeiensis f. *pteracantha*, 201–202

Ruscus aculeatus, 23, 203–205

Sansevieria cylindrica, 272–273

Sansevieria trifasciata, 273

Sarracenia, 59–63

   S. flava, 33, 62

   S. leucophylla, 61

   S. purpurea, 61

Sauromatum guttatum, 137

scouring-rush, 77–78

sea holly, 96–97

sea urchin cactus, 249

semaphore plant, 157

Senecio rowleyanus, 262

sensitive mimosa, 160

sensitive plant, 160–162

shield flower, 141

shield sundew, 50

Sinningia leucotricha, 24

snake lily, 137

snake plant, 273

snow-on-the-mountain, 144

Solanaceae, 186–187

Solanum mammosum, 186–187

Solanum pyracanthum, 206–207

Solanum quitoense, 207

South American pitcher plant, 51

Spanish moss, 165

spathe and spadix, 89, 91–92, 140, 164, 136

spear sansevieria, 272–273

sphagnum moss, 61, 165

spider orchid, 215–217

spider's tresses, 110–112

spurge, 143

staghorn fern, 74–76

Stapelia gigantea, 257

Stapelia hirsuta, 256–257

stapeliad, 256–258

starfish flower, 256

Strelitzia alba, 189

Strelitzia nicolai, 189

Strelitzia reginae, 188–189

string-of-hearts, 129–131

Strophanthus preussii, 110–112

sundew, 48–50

Tacca chantrieri, 113–115, 166–167

Tacca integrifolia, 113–115

tapeworm plant, 178–179

tarantula cactus, 174–175

tasselvine, 111

telegraph plant, 157–159

tepuis, 51

Tillandsia, 163–165

Tillandsia usneoides, 165

titan arum, 15, 87–90

tongue orchid, 220

trash basket roots, 21

trifoliate orange, 199

trigger orchid, 223–225

Triplochlamys multiflora. See *Pavonia* × *gledhillii*

tropical pitcher plant, 54–56

unicorn plant, 184–185

Venus flytrap, 44–47

voodoo lily, 135–135

water fern, 71

wax flower, 25–26

wax hearts, 115–117, 129–131

western Australian pitcher plant, 38–40

whisk fern, 82–83

white bat plant, 113

white jack-in-the-pulpit, 140

wild bleeding heart, 126

wild ginger, 169

window-leaved succulents, 260–262

wingthorn rose, 201–202

Xanthopan morganii praedicta, 213

Zingiber spectabile, 148–149